Single-Photon Avalanche Diodes and Photon Counting Systems

Marc Dandin • Nicole McFarlane •
Md Sakibur Sajal • Fahimeh Dehghandehnavi •
Babak Nouri

Single-Photon Avalanche Diodes and Photon Counting Systems

From Phototransduction to Circuit Architecture

Springer

Marc Dandin
Carnegie Mellon University
Pittsburgh, PA, USA

Nicole McFarlane
University of Tennessee at Knoxville
Knoxville, TN, USA

Md Sakibur Sajal
Carnegie Mellon University
Pittsburgh, PA, USA

Fahimeh Dehghandehnavi
Carnegie Mellon University
Pittsburgh, PA, USA

Babak Nouri
Virginia Tech
Blacksburg, VA, USA

ISBN 978-3-031-64333-0 ISBN 978-3-031-64334-7 (eBook)
https://doi.org/10.1007/978-3-031-64334-7

This Springer imprint is published by the registered company Springer Nature Switzerland AG
The registered company address is: Gewerbestrasse 11, 6330 Cham, Switzerland

To Anayiza, Parvin, Hossein, Lyla, Rimi, Maliha, Bari, Iza, Noa, Ari, Anya, Sanite, Marvel, and Yanick.

Preface

There has been significant progress in the development of single-photon avalanche diodes (SPADs) since their inception in the 1970s, yet their use in mainstream applications has lagged behind. In fact, as of the writing of this manuscript, manufacturers of traditional image sensors are only timidly exploring the commercial space for SPADs. Therefore, it is safe to say that despite the advances we have witnessed in the past decades, SPAD technology still has much to prove. Particularly, their integration in complementary metal-oxide semiconductor (CMOS) processes is still an active area of research with many unmet challenges.

Our aim with the present monograph is to contribute to that specific body of knowledge. Therefore, our discussion is restricted solely to CMOS SPADs and to their associated readout and processing circuits. The monograph's scope mirrors that of a tutorial we delivered at the 29th IEEE International Conference on Electronics, Circuits, and Systems (ICECS 2022) held in Glasgow, UK, in December 2022. There, we presented the current state of CMOS SPAD research, and we particularly focused on our own efforts developing perimeter-gated SPADs (pg-SPADs) in standard CMOS technologies. This experience catalyzed this book project, and we are delighted to finally present a monograph that covers SPAD technology from the bottom-up, i.e., starting from fundamental topics like phototransduction and reaching advanced topics like silicon photomultiplier (SiPM) architectures and CMOS imaging with pg-SPADs.

Our target audience includes the curious engineer who has no background in SPAD technology. It also includes SPAD researchers who are eager to explore novel ideas in this field. Each of the book's chapters may be read individually, providing an in-depth view about a particular aspect of CMOS SPAD technology.

Chapter 1 briefly discusses the fundamental notions underpinning photon detection in semiconductor devices. It covers photo-electron emission in semiconductors, noise, and typical photosensor front-end architectures, including Geiger mode front-ends, which include SPAD devices. Chapter 2 covers the issue of premature edge breakdown in SPADs implemented CMOS technologies, and it presents perimeter gating as a viable alternative to mitigate this issue. Chapter 3 focuses on the optoelectronic characterization of CMOS SPADs, covering instrumentation and measurement strategies. Chapter 4 covers the implementation of a perimeter-gated SPAD imager in a standard CMOS technology. Chapters 5 covers a novel application for SPADs, namely their use in hardware security applications. Chapter 6 discusses silicon photomultipliers, and Chap. 7 reviews readout architectures. Chapter 8 covers dead time correction models. We conclude the book in Chap. 9, summarizing key findings and briefly describing the foreseeable future of pg-SPAD research and development. All throughout, the reader will find a myriad of topics relating to SPADs and photon counting systems in general.

In closing, we hope that the reader will truly enjoy this book and its fresh take on this emerging field. Lastly, we thank our numerous colleagues and students for contributing to this material; they are acknowledged more appropriately in the following *Acknowledgments* section.

Pittsburgh, PA, USA — Marc Dandin
Knoxville, TN, USA — Nicole McFarlane
May 2024

Acknowledgements

We are indebted to our colleagues and mentors, first, for being a constant source of inspiration, and second, for supporting our interest in optical sensor and microsystems research for nearly two decades. We thank the following individuals for their unwavering support at various points in our journey and to this day. They are: Dr. Larry Pierre, Dr. Pamela Abshire, Dr. Elisabeth Smela, Dr. Robert Fischell, Dr. Honghao Ji, Dr. Akin Akturk, Dr. Valeri Saveliev, Dr. Bathiya Senevirathna, Dr. David Sander, Dr. Kai-Chun Lin, Mr. Sheung Lu, Dr. Khandaker Mamun, Dr. Md. Habib Habib, Dr. Shamim Shawkat, and Mr. Jinlong Gu.

We are also grateful to our students, past and present. We trust that their unfettered enthusiasm and their steadfast engagement will lead to novel avenues in SPAD research, and we are excited to see their growth as the next generation of researchers in this field. We offer special thanks to the students who currently spearhead SPAD efforts in our labs: Mr. Md. Sakibur Sajal, Ms. Fahimeh Dehghandehnavi, Mr. Atik Rahman, Mr. Andalib Nizam, and Ms. Farin Rahman. Generally, we offer our thanks to all the members of our respective research groups, the Integrated Circuits and Bioengineering Laboratory (ICBio Lab) at Carnegie Mellon University, and the M-Lab at the University of Tennessee, Knoxville. We thank them for their intellectual contributions to the many research projects in our portfolios and for their continued trust in our leadership.

We also kindly acknowledge the support of our research administrators without whom none of this work would be possible; our special thanks go to Ms. Jennifer Agnew, Mr. Kevin Bosle, Ms. Allison Ervin, and to Ms. Kristen Geiger and her team, for their support of the ICBio Lab. Lastly, we thank our sponsors for supporting our research and educational endeavors. We particularly acknowledge the United States Department of Energy, the United States National Science Foundation, and Cylab at Carnegie Mellon University for their support.

Contents

Acronyms

ADC	Analog-to-digital converter
AER	Address event representation
APD	Avalanche photodiode
APS	Active pixel sensor
AQAR	Active quench active reset
BiCMOS	Bipolar complementary metal-oxide semiconductor
BJT	Bipolar junction transistor
CCD	Charge-coupled device
CFD	Constant fraction discriminator
CFPN	Column fixed-pattern noise
CMFB	Common mode feedback
CMOS	Complementary metal-oxide semiconductor
CRP	Challenge-response pair
CSP	Charge-sensitive pre-amplifier
CTR	Coincidence time resolution
DAQ	Data acquisition
DC	Direct current
DCM	Dark count map
DCR	Dark count rate
DCT	Discrete cosine transform
DNF	Denoised figure
DSNU	Dark signal non-uniformity
DUT	Device under test
EMVA	European Machine Vision Association
EVG	Event generator
FLIM	Fluorescence lifetime imaging microscopy
FPGA	Field-programmable gate array
FPN	Fixed-pattern noise
FWHM	Full-width half-max
GMAPD	Geiger mode avalanche photodiode
HD	Hamming distance

ICECS	(IEEE) International Conference on Electronics, Circuits, and Systems
IOT	Internet-of-things
JFET	Junction field-effect transistor
KDC	Key distribution center
LED	Leading edge discriminator
LIDAR	Light detection and ranging
LOC	Lab-on-a-chip
LYSO	Lutetium–yttrium oxyorthosilicate
MCLMS	Multichannel least mean square
MEMS	Micro-electro-mechanical systems
MOSFET	Metal-oxide field-effect transistor
MSE	Mean square error
NEP	Noise-equivalent power
NHD	Normalized Hamming Distance
NIST	National Institute of Standards and Technology
PCB	Printed circuit board
PD	Photo detector
PDE	Photo detection efficiency
PEB	Premature edge breakdown
PET	Positron emission tomography
PG-SPAD	Perimeter-gated single-photon avalanche diodes
PMT	Photo-multiplier tube
PRNG	Pseudo-random number generator
PRNU	Photoresponse non-uniformity
PUF	Physically unclonable function
PZC	Pole zero cancellation
QLC	Quenching logic circuit
RBE	Random bit efficiency
RLC	Reset logic circuit
RNG	Random number generator
ROC	Readout circuit
SAR	Successive approximation register (ADC)
SCD	Symmetric charge division multiplexing
SIPM	Silicon photomultiplier
SNR	Signal-to-noise ratio
SPAD	Single-photon avalanche diode
STI	Shallow trench isolation
STS	Statistical test suite
SMU	Source measure unit
TAC	Time-to-amplitude converter
TDC	Time-to-digital converter
TIA	Transimpedance amplifier
TOT	Time-over-threshold
TRNG	True random number generator
VDE	Variable delay element

Chapter 1
Fundamentals of Phototransduction in Semiconductors

1.1 Introduction

The study of the interaction of light with semiconductors is extensive; in this chapter we provide only the fundamental notions that are necessary to understand the operation of a photosensor.[1] Our discussion begins with the physical mechanism underlying the conversion of photons to free carriers in a semiconductor material. This process, which is referred to as photogeneration or photoelectric emission, is at the core of imaging science, and a thorough understanding of its nature is required when designing the front end and the readout chain [1].

We then center our discussion on the concept of phototransduction and on a set of system-level performance metrics, namely, the SNR, the noise equivalent power (NEP), and the spectral responsivity ($S(\lambda)$). Taken together, these figures of merit provide important insights into device performance by allowing the identification, at the design stage, of the trade-offs that exist between various circuit parameters. To illustrate this fact and also to understand the operation of the SPADs and pg-SPADs featured in this book, we focus on three distinct photosensor configurations, namely, the photodiode, the avalanche photodiode, and the single-photon avalanche photodiode configuration.

[1] Here, the term photosensor is used to designate the front-end (i.e., the detector) and the readout circuit (ROC).

M. Dandin et al., *Single-Photon Avalanche Diodes and Photon Counting Systems*,
https://doi.org/10.1007/978-3-031-64334-7_1

1.2 Photoelectron Emission in Semiconductors

1.2.1 Absorption Coefficient

The transmission of light through a semiconductor is governed by the Beer-Lambert law (shown below) [2]. It states that within the material, an incident light beam (of intensity I_{IN}) is attenuated by an exponential factor that depends on the material's thickness (d) and on a wavelength-dependent parameter α, the absorption coefficient. Thus, the exiting light beam (of intensity I_{OUT}) has a smaller intensity than the incident beam. Figure 1.1 shows the absorption coefficient data for commonly used semiconductors.

$$I_{OUT} = I_{IN} e^{-\alpha d}. \tag{1.1}$$

If the incident photons have enough energy, they will cause the valence electrons to migrate to the conduction band. These band-to-band transitions occur only when the incoming photons have energies that exceed the band gap energy of the semiconductor [2]; otherwise, the photons pass through the material.

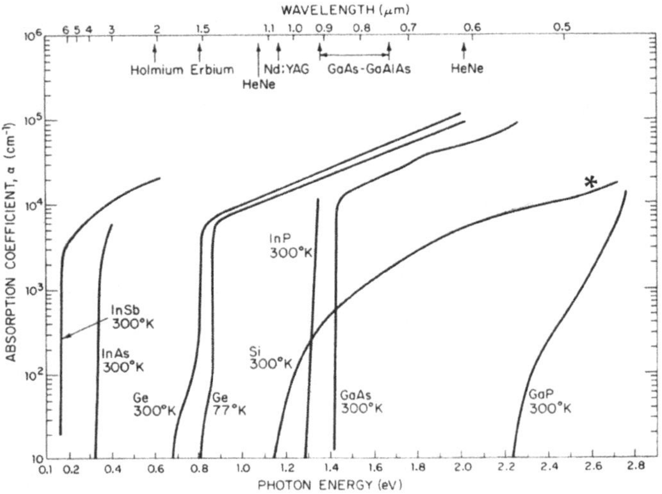

Fig. 1.1 Absorption coefficient of different semiconductor materials as a function of photon energy and wavelength. The asterisk indicates the data for silicon, the material with which the detectors featured in this book were fabricated. (Figure adapted from ref. [2])

1.2.2 Quantum Efficiency

Once photons are absorbed, they generate charge with an efficiency factor denoted η. This *quantum* efficiency is an important metric in detector design. It is the ratio of the number of photoelectrons generated to the number of photons incident on the material. Since photon absorption is wavelength-dependent as shown above, so is η.[2] Moreover, it also depends on the reflection coefficient (R) of the interface formed by the medium in which the incident light originates and the semiconductor's surface [4].

1.2.3 Noise

Here we cursorily introduce the concept of noise. We refer to noise as the sum of the effects of secondary physical processes that contribute to the detector's signal, the primary process here being the absorption and transduction of the photons of interest. Such secondary processes may include the thermal generation of carriers, band-to-band tunneling, or releases from trapped carriers. There are also sources of noise in the readout chain, and these do also contribute to the overall noise associated with the phototransduction process. Furthermore, the uncertainty in the impinging light itself may give rise to a noisy signal; this process is termed *shot noise*, and it is described in detail below.

For ease of analysis, we focus only on the detector front end for now, i.e., we do not consider readout noise processes. Let us first consider how carriers are generated in the detector during phototransduction; this happens via photoelectric emission in which electrons jump from the valence to the conduction band.

An ensemble measurement of the number of photogenerated carriers will show a spread around a mean value. This uncertainty is the aggregate result of several noise phenomena originating from one or more of the processes mentioned above. In a traditional detector, the two most important source of noise can be the inherent randomness in the impinging light quanta, i.e., the shot noise, and the random motion of charges due to thermal energy [4].

While the latter effect can be mitigated by cooling the detector and actively regulating the ambient temperature, the former cannot be suppressed. Shot noise is the fundamental noise limit of photodetection, and it sets a lower bound to the noise figure of optical sensing circuits [3, 4]. Because of its importance, we spend some

[2] If the incoming light is modulated, the quantum efficiency also depends on the modulation frequency. This dependence is not fundamental; rather it is related to circuit and device parameters (see [3]). Thus, for completeness, the quantum efficiency is denoted $\eta(\lambda, w)$, where λ is the photon's wavelength and w the modulation frequency.

time in the following section analyzing its origin, and we formulate a relationship for it, which we will use in our front-end models in the remainder of this chapter. The analysis featured is credited to Davenport and Root [5], and although a vacuum diode was used as a model for arriving at the result, the analysis extends well to all photosensor front ends.

1.2.3.1 Shot Noise

In an optical sensor front end, the photogenerated current exhibits random fluctuations that are independent of the ROC noise sources and of thermal energy. These fluctuations are known as the shot noise, and they arise from the discrete and random removal of energy quanta from the incident radiation field. For this reason, shot noise is a quantum mechanical random process, and it is the fundamental noise which underlies all light detection and measurement applications.

In a detector that integrates photogenerated charge, like an active pixel sensor (APS), for example, the shot noise manifests itself as an uncertainty in the source follower-buffered voltage of the integration node. On the other hand, in a detector that performs photoelectric counting, like the pg-SPAD, it can be estimated from the uncertainty in the count rate.

Thus, the number of photoelectrons generated in a detector is a random process, irrespective of the detector's mode of operation and architecture. The distribution that describes this random process can be derived by considering photoelectron emission in an ideal vacuum photocathode (shown in Fig. 1.2).

1.2.3.2 Single-Electron Current in a Vacuum Diode

In Fig. 1.2a the plate (or anode) is biased at a constant voltage denoted V_A and the photocathode is held at ground. This system is governed by the following electrostatic and classical kinematic equations:

$$\mathbf{F}_m = m\mathbf{a} \tag{1.2}$$

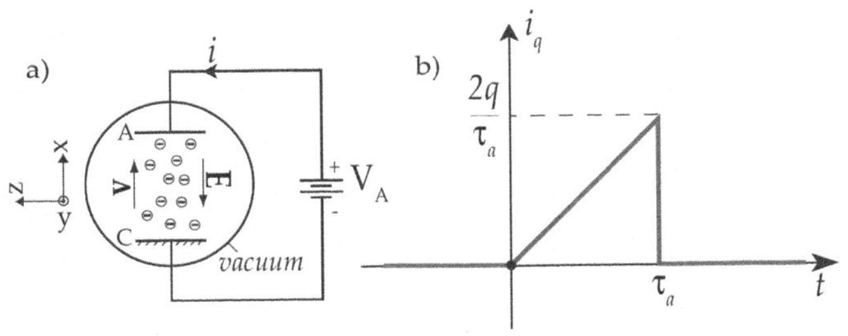

Fig. 1.2 (**a**) Vacuum photodiode. (**b**) Single-electron current pulse

$$\mathbf{F}_l = -q\mathbf{E} \tag{1.3}$$

$$\mathbf{E} = -\nabla V \tag{1.4}$$

$$\nabla^2 V = -\frac{\rho}{\epsilon} \tag{1.5}$$

$$\mathbf{J} = \rho \boldsymbol{v} \tag{1.6}$$

Here, $\mathbf{F}_m$ is the Newtonian force that a moving electron is subjected to as a result of its acceleration through the electric field $\boldsymbol{E}$ resulting from the potential difference between the two electrodes, ρ is the charge density, $\boldsymbol{J}$ the current density, and $\boldsymbol{v}$ the velocity of the electron. $\mathbf{F}_l$ is the Lorentz force in the absence of a magnetic field ($\boldsymbol{B} = 0$).

We first solve for the potential as a function of distance. We restrict our analysis to only one dimension, that is, the y and z components of the E-field are zero. We further assume that temperature is constant and that there are no space charge effects. The latter assumption means that the electron does not interact with other electrons during its motion and that its initial velocity, in comparison to its terminal velocity (at the plate), is negligible. Consequently, Poisson's equation (1.5) simplifies to Laplace's equation $\left(\nabla^2 V = 0\right)$. This yields the boundary value problem (Eqs. 1.7, 1.8, and 1.9) whose solution is shown in Eq. 1.10:

$$\frac{\delta^2 V}{\delta x^2} = 0 \tag{1.7}$$

$$V(0) = 0 \tag{1.8}$$

$$V(d) = V_A \tag{1.9}$$

$$V(x) = \frac{V_A}{d} x \tag{1.10}$$

The electric field is obtained using Eq. 1.4. In Eq. 1.11, the vector $\hat{\mathbf{i}}$ indicates the basis vector in the x-direction in a Cartesian reference frame:

$$\mathbf{E} = -\frac{V_A}{d}\hat{\mathbf{i}} \tag{1.11}$$

We may now find a second-order differential that describes the motion of the electron by using a force balance equation. This is done by equating the Lorentz force to the Newtonian force. This yields

$$\mathbf{F}_l = \mathbf{F}_m \tag{1.12}$$

$$q\frac{V_A}{d}\hat{\imath} = m\frac{d^2x}{dt^2}\hat{\imath} \tag{1.13}$$

$$q\frac{V_A}{dm} = \frac{d^2x}{dt^2} \tag{1.14}$$

By integrating Eq. 1.14 once, we obtain the velocity. Recall that we assume the initial velocity of the electron to be negligible compared to terminal velocity; this means v(0) = 0. Further integration yields a closed-form solution (Eq. 1.16) for the position of the electron as it travels towards the plate:

$$\mathbf{v(t)} = \frac{qV_A}{dm}t \tag{1.15}$$

$$\mathbf{x(t)} = \frac{1}{2}\frac{qV_A}{dm}t^2 \tag{1.16}$$

The transit time, i.e., the time it takes the electron to travel a distance equal to d, is found by solving for t in Eq. 1.16:

$$\tau_a = \sqrt{\frac{2m}{qV_A}d} \tag{1.17}$$

We now have all the equations that are needed in order to find the current induced on the plate as a result of the motion of the single charge. This is done by setting up an energy balance equation that relates the potential energy of the moving electron and the work that needs to be done to induce a net charge Q at the plate:

$$U = q\frac{V_A}{d}x \tag{1.18}$$

$$W = QV_A \tag{1.19}$$

$$q\frac{V_A}{d}x = QV_A \tag{1.20}$$

$$i_e(t) = \frac{dQ}{dt} \tag{1.21}$$

$$i_e(t) = \frac{q}{d}\frac{dx}{dt} \tag{1.22}$$

Expressing Eq. 1.22 in terms of the transit time, we obtain the final expression for the current pulse resulting from pulling a single charge from the cathode, and it is plotted in Fig. 1.2b:

$$i_e(t) = \left(\frac{2q}{\tau_a}\right)\left(\frac{t}{\tau_a}\right), \quad \text{for } 0 \le t \le \tau_a \tag{1.23}$$

1.2.3.3 Photoelectron Emission Probability in a Time Interval τ

Equation 1.23 suggests that each electron emitted from the photo-cathode creates a current pulse of duration τ_a. In this section, we derive the statistical distribution that accounts for the number of these current pulses in a time interval of duration τ, where $\tau \ge \tau_a$. We begin by deriving the probability that no pulses are emitted in a time window τ; using that result, we subsequently derive the probability of having k events in that same time interval.

In order to arrive at the latter, we assume that, in an incremental increase $\Delta\tau$, there can either be only one event, or no event at all. Mathematically, this is written as shown in Eq. 1.24, with p denoting the probability measure:

$$p(0, \Delta\tau) + p(1, \Delta\tau) = 1 \tag{1.24}$$

We now assume that the probability of obtaining one event in the incremental time window is proportional (by a constant) to the length of the time window. That is,

$$p(1, \Delta\tau) = a\Delta\tau \tag{1.25}$$

If we further assume that the emission of one electron is independent of previous emissions, then statistical independence allows us to write

$$p(0, \tau + \Delta\tau) = p(0, \tau)p(0, \Delta\tau) \tag{1.26}$$

Substituting in Eq. 1.25 and Eq. 1.24 in Eq. 1.26, we find an expression for $p(0, \Delta\tau)$. Taking the limit of $\Delta\tau \to 0$ in Eq. 1.27 yields a first-order difference equation whose solution is shown in Eq. 1.28. The difference equation was solved using the boundary condition shown in Eq. 1.29, which implies that when the time window $\Delta\tau$ is vanishingly small, there can be no emissions. In other words, the probability of having no emissions in a zero-width time interval is equal to 1:

$$\frac{p(0, \tau + \Delta\tau) - p(0, \tau)}{\Delta\tau} = -ap(0, \tau) \tag{1.27}$$

$$p(0, \tau) = e^{-a\tau} \tag{1.28}$$

$$p(0, 0) = \lim_{\Delta\tau \to 0} p(0, \Delta\tau) = 1 \tag{1.29}$$

We now derive the probability of having K events in a time interval of duration τ. This probability, denoted $P(K, \tau)$, is a general expression for calculating the likelihood of having an arbitrary number of electron emissions in a given time window. The derivation begins by considering a time interval of duration $\tau + \Delta\tau$ and by again assuming that emissions times are statistically independent.

We write the following, with the latter assumption permitting the expansion of the left-hand side of Eq. 1.30 as a product of probabilities:

$$p(k, \tau + \Delta\tau) = p(k, \tau) \times p(k, \Delta\tau) \tag{1.30}$$

However, it is more convenient to expand the left-hand side of Eq. 1.30 as a sum of conditional probabilities by virtue of the assumed statistical independence of the emission times. Equation 1.30 becomes

$$p(k, \tau + \Delta\tau) = p(k - 1, \tau; 1, \Delta\tau) + p(k, \tau; 0, \Delta\tau) \tag{1.31}$$

Furthermore,

$$p(k, \tau + \Delta\tau) = p(k - 1, \tau)p(1, \Delta\tau) + p(k, \tau)p(0, \Delta\tau) \tag{1.32}$$

Substituting in the values previously derived for $P(0, \tau)$ and $P(1, \tau)$ into Eq. 1.32 and dividing through by $\Delta\tau$ yields

$$\frac{p(k, \tau + \Delta\tau) + p(k, \tau)}{\Delta\tau} + ap(k, \tau) = p(k - 1, \tau) \tag{1.33}$$

For $\Delta\tau \rightarrow 0$, Eq. 1.33 becomes a recursive differential equation which establishes a relation between $p(k, \tau)$ and $p(k - 1, \tau)$. Utilizing an appropriate integrating factor and the initial value $p(0, 0) = 0$, the solution of Eq. 1.33 is

$$p(k, \tau) = ae^{-a\tau} \int_0^{\tau} e^{at} p(k - 1, t)dt \tag{1.34}$$

Using mathematical induction and $p(0, \tau) = e^{-a\tau}$, the following general expression is obtained:

$$p(k, \tau) = \frac{(a\tau)^k e^{-a\tau}}{k!} = \frac{(n)^k e^{-n}}{k!} \tag{1.35}$$

The above equation is none other than the well-known Poisson probability mass function, $p(k, n)$, which describes arrival processes with mean $n = a\tau$. As expected, the normalization condition $\left(\sum_{k=0}^{\infty} p(k, n) = 1\right)$ is satisfied, and for large values of k, the limit is Gaussian.

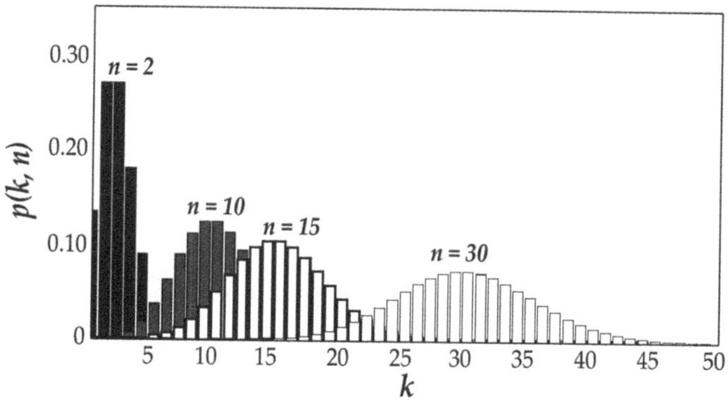

Fig. 1.3 Poisson distribution for $n = 2, 10, 15, 30$

Equation 1.35 is plotted in Fig. 1.3 for various n, and it implies that the current I through the vacuum diode during time τ is proportional to the number of k electrons emitted at a fixed rate $r = \frac{k}{\tau}$ (Eq. 1.36). However, the mean square current is given by Eq. 1.37:

$$I = \frac{kq}{\tau} \tag{1.36}$$

$$\left\langle \Delta I^2 \right\rangle = \left\langle I^2 \right\rangle - \langle I \rangle^2 \tag{1.37}$$

The first term in the right-hand side of Eq. 1.37 is computed directly from Eq. 1.36:

$$\left\langle I^2 \right\rangle = \left\langle k^2 \right\rangle \left(\frac{q}{\tau} \right)^2 \tag{1.38}$$

Similarly, the second term is

$$\langle I \rangle^2 = \langle k \rangle^2 \left(\frac{q}{\tau} \right)^2 \tag{1.39}$$

Substituting Eqs. 1.38 and 1.39 into Eq. 1.37 yields

$$\left\langle \Delta I^2 \right\rangle = \left(\frac{q}{\tau} \right)^2 \left(\left\langle k^2 \right\rangle - \langle k \rangle^2 \right) \tag{1.40}$$

The second factor in Eq. 1.40 can be rewritten as[3]

$$\left(\left\langle k^2\right\rangle - \langle k\rangle^2\right) = \left(\left\langle k^2\right\rangle - n^2\right) = n \tag{1.41}$$

Thus, Eq. 1.40 yields

$$\left\langle \Delta I^2\right\rangle = n\left(\frac{q}{\tau}\right)^2 = \left(\frac{\langle I\rangle q}{\tau}\right) \tag{1.42}$$

Rewriting $\left\langle \Delta I^2\right\rangle$ as $\overline{i^2}$ and $\frac{1}{\tau}$ as 2β, where β is the single-sided measurement bandwidth, yields the well-known equation for shot noise:

$$\overline{i^2} = 2q\langle I\rangle\beta \tag{1.43}$$

1.3 Phototransduction and System-Level Metrics

Phototransduction, although subtly related to photoelectron emission, is functionally different. It is the process through which photogenerated carriers are converted into an electrical signal (i.e., a current or a voltage) that can be processed to infer, using a calibration curve, radiometric (e.g., light intensity) or photoelectric (e.g., photon flux) characteristics of the incident light beam. It is thus a system-level process unlike photoelectric emission. The metric that quantifies its performance is the spectral responsivity, which depends, at some level, on the quantum efficiency, but is mainly influenced by the detector's biasing and the properties of its ROC.

The following sections present an analytical discussion of three photosensor front ends. They are the photodiode (PD), the avalanche photodiode (APD), and Geiger-mode avalanche photodiode (GMAPD), the latter being the central focus of this book. All three comprise a p-n junction, but in each case, the junction is biased in a distinct reverse bias regime (Fig. 1.4).

Since deriving the spectral responsivity analytically requires a device physics-based model that takes into account dimensions as well as material properties, the discussion focuses rather on the signal-to-noise ratio (SNR) metric which can be computed from noise-equivalent circuit models; that is the case the first two front ends discussed. In case of the third one, i.e., the GMAPD, responsivity as well as SNR is easily determined experimentally. Techniques for making these measurements are discussed later throughout the book.

[3] This is done by subtracting the first and second central moments of Eq. 1.37 and subsequently making use of the normalization condition [4].

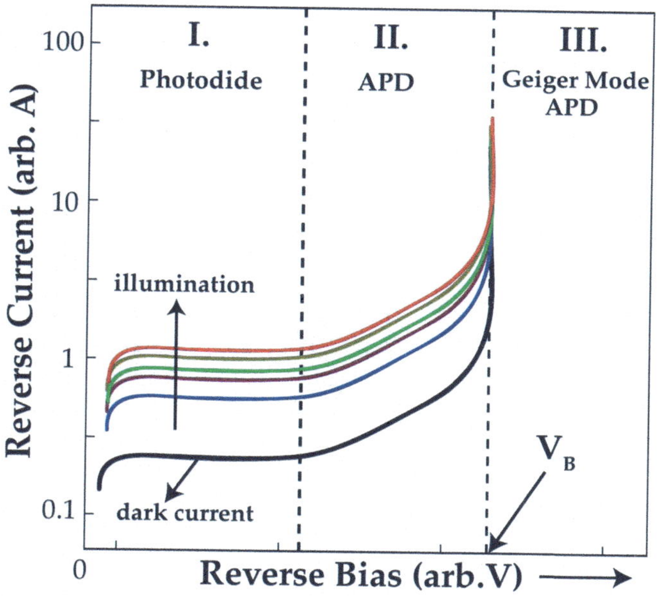

Fig. 1.4 Arbitrary I-V curve showing three distinct p-n junction biasing regimes. (Figure adapted from ref. [6])

1.4 Photosensor Architectures

1.4.1 Photodiode Front Ends

Photodiodes are p-n junctions biased at low reverse biases (Region I in Fig. 1.4). In that regime, at most one free-carrier pair can be generated from an absorbed photon, that is, there is no internal gain mechanism through which a photogenerated carrier can create secondary electron-hole pairs. As shall be seen later on, this is not the case in Regions II and III.

Since the detector is biased, photogenerated carriers are accelerated by the applied electric field, and this gives rise to a photocurrent. As an ROC, for example, a transimpedance preamplifier may be used to convert the photocurrent into a voltage for further processing. A generic representation of a photodiode-based photosensor is shown in Fig. 1.5.

1.4.1.1 Photodiode Signal-to-Noise Ratio

This section analyzes the photodiode's SNR utilizing the noise-equivalent circuit model shown in Fig. 1.5. The derivation of the SNR metric from the photodiode noise-equivalent circuit model, as shown below, is credited to Stillman and Wolfe [3].

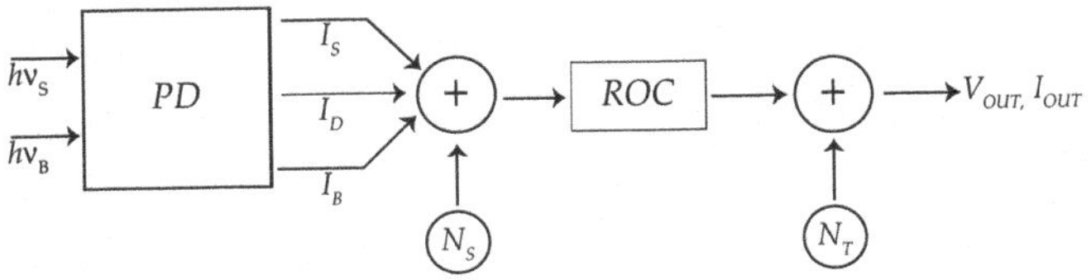

Fig. 1.5 Model of a photodiode (PD) with its readout circuit (ROC). h is Planck's constant and ν the frequency of the light. N_S and N_T denote shot noise and thermal noise terms, respectively. (Figure adapted from ref. [3])

The light incident on the junction has two components: one, of energy $h\nu_s$, which is the signal of interest, and the other, a background light signal of energy $h\nu_b$. Both components of the incoming light signal contribute to the photocurrent. In Fig. 1.5, the mean values of these current components are denoted I_S and I_B, respectively.

Moreover, in the absence of light, a current can still be measured through the photodiode. That current is called the dark current, and it has a mean value denoted I_D. It is due to thermal energy in the junction and other leakage mechanisms from neighboring circuit components.

The dominant noise source in the ROC is thermal noise, and it can be modeled in its entirety as a two-port resistive impedance which has thermal noise current given by Eq. 1.44 [7]:

$$\overline{i_n^2} = 4\mathrm{kT/R} \tag{1.44}$$

Accordingly, and utilizing the noise-equivalent circuit model shown in Eq. 1.44, an expression for the SNR can be written as the ratio of the root-mean-square (rms) signal power to that of the sum of all the noise contributions [3]. These quantities are determined below. Furthermore, the input signal power is assumed to be modulated as shown in Eq. 1.45.

Here, the modulation factor is m, the signal amplitude P_0, and the angular frequency w. The root-mean-square signal power is shown in Eq. 1.46:

$$P(wt) = P_0(1 + m\cos(wt)) \tag{1.45}$$

$$\langle P \rangle^2 = \frac{mP_0}{\sqrt{2}} \tag{1.46}$$

The rms signal current is thus given by

$$i_P(\lambda, w) = q\eta(\lambda, w)\frac{mP_0}{h\nu_S\sqrt{2}} \tag{1.47}$$

The average photocurrent[4] is given by

$$I_P(\lambda, 0) = q\eta(\lambda, 0)\frac{mP_0}{h\nu_S\sqrt{2}} \tag{1.48}$$

The average background current, on the other hand, is the current produced over all wavelengths of light incident on the device. This is expressed mathematically as

$$I_B = q\int_\lambda \frac{\left[\eta\left(\lambda', 0\right) P_B\left(\lambda'\right)\right]}{h\nu'} d\lambda' \tag{1.49}$$

The resistances in the equivalent circuit shown in Fig. 1.6 can be lumped into one resistance R_{eq} at an effective temperature denoted by T_{eff}.

The SNR is

$$\left(\frac{S}{N}\right) = \frac{\frac{1}{2}\left(\frac{q\eta P_0}{h\nu_S}\right)^2}{2q\left(I_P + I_B + I_D\right)\beta + \left(4kT_{eff}\left(\frac{1}{R_{eq}}\right)\beta\right)} \tag{1.50}$$

Note that in the denominator, the first term of the sum is the additive shot noise obtained from Eq. 1.43, while the second term is the mean-squared thermal noise. To analyze the photodiode's performance, it is more useful to solve Eq. 1.50 for the optical power P_0 as a function of $\left(\frac{S}{N}\right)$. This yields

$$P_0 = \frac{2h\nu_S\beta}{\eta}\left(\frac{S}{N}\right)\left\{1 + \left[1 + \frac{I_{eq}}{q\beta\left(\frac{S}{N}\right)}\right]^{\frac{1}{2}}\right\} \tag{1.51}$$

where

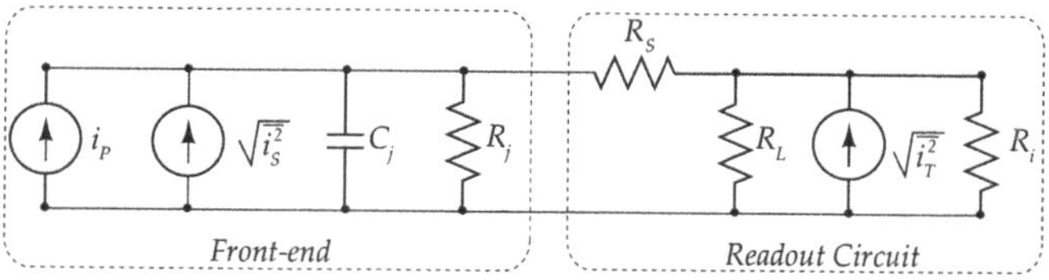

Fig. 1.6 Noise equivalent small signal model of a photodiode with ROC. i_p is the root-mean-square photocurrent, while the other two current sources represent the mean-square shot noise and thermal noise terms. (Figure adapted from [3])

[4] Here, we have changed the subscript from S to P to indicate that this component is due to the signal of amplitude P_0.

$$I_{eq} = I_B + I_D + \left(\frac{2kT_{eff}}{qR_{eq}}\right) \tag{1.52}$$

Equation 1.51 can be analyzed in two limiting regimes. In the first regime, the second term in the square brackets vanishes if the bandwidth is large. In that situation, the minimum power required to achieve a given SNR is limited only by the quantum efficiency. In reality, however, attaining such a limit is not possible because widening the bandwidth increases the thermal noise in the ROC (*i.e.*, T_{eff} increases and R_{eq} decreases in Eq. 1.52); this means the term $\frac{I_{eq}}{q\beta\left(\frac{S}{N}\right)}$ may not go to zero.

In the other limiting case, the following condition holds:

$$\frac{I_{eq}}{q\beta\left(\frac{S}{N}\right)} \gg 1 \tag{1.53}$$

In this regime, design trade-offs are best understood utilizing the noise-equivalent power (NEP) metric, which is the optical power that is required to obtain an SNR of 1 in a 1 − Hz measurement bandwidth [4]. A high-performance detector has a low NEP. The NEP is found from Eq. 1.51 by setting the SNR term to 1 and by applying the limit condition of Eq. 1.53 for further simplification:

$$NEP = \sqrt{2}\left(\frac{h\nu}{\eta}\right)\left[\frac{I_{eq}}{q}\right]^{\frac{1}{2}} = \sqrt{2}\left(\frac{h\nu}{\eta}\right)\left[\frac{I_B + I_D + \left(\frac{2kT_{eff}}{qR_{eq}}\right)}{q}\right]^{\frac{1}{2}} \tag{1.54}$$

As Eq. 1.54 shows, the NEP is minimized when R_{eq} is large. Unfortunately, however, large values of R_{eq} significantly reduce the 3 dB cutoff frequency of the ROC and thus lowers the bandwidth. In conclusion, the photodiode front end, equipped with a generic ROC, is best suited for applications that are background-limited (i.e., large I_B) and require small bandwidths [3]. As stated above, that limit can be reached by increasing R_{eq} at the expense of the bandwidth.

1.4.2 Avalanche Photodiode Front Ends

This section considers a photosensor whose front end is biased in Region II of Fig. 1.4. In that regime, the front end is said to be an avalanche photodiode. The applied field is so strong that it imparts enough kinetic energy to the photogenerated carriers; upon impacting a lattice atom as they are accelerated, they can create secondary carriers. These secondary carriers are in turn accelerated and give rise to additional free carriers via the same mechanism. This phenomenon is termed impact ionization, and it gives rise to an avalanche current, hence the APD denomination.

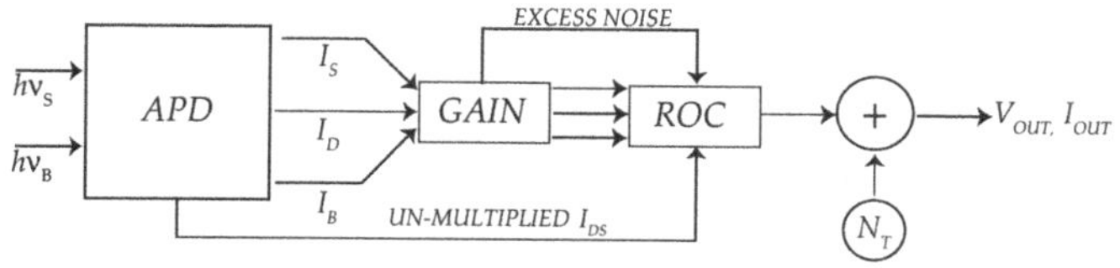

Fig. 1.7 Model of an APD with ROC. The junction now has an internal gain mechanism which multiples the mean currents as well as the noise. Some of the dark current remains un-multiplied and the thermal noise term of the ROC is the same as in the case of the PD with ROC. (Figure adapted from [3])

In Region II, the avalanche photocurrent is proportional to the photocurrent that would be generated without avalanche. The proportionality factor is finite, and it is called the multiplication factor, denoted M. This parameter is not only wavelength-dependent but it is also influenced by the modulation frequency of the incident light signal [3].

A model of the APD with ROC is shown in Fig. 1.7. Whereas the photocurrent is multiplied, the noise terms are also amplified. Intuitively, it would seem that an APD has a lot more noise than a photodiode front end. However, there are applications for which an APD is preferable to a photodiode because much lower NEPs can be achieved. As shown below, deriving the SNR and NEP metrics readily identifies the design trade-offs of the APD front end. This analysis is also credited to Stillman and Wolfe [3].

1.4.2.1 Avalanche Photodiode Signal-to-Noise Ratio

The noise-equivalent circuit model for the APD-based photosensor is the same as that of the photodiode-based sensor (shown in Fig. 1.5). As before, shot noise, thermal noise, background, signal of interest, and dark current all couple to the ROC. The difference for an APD, however, is that, barring thermal noise in the ROC, all these components will be amplified by the internal gain mechanism. Moreover, statistical fluctuations in the gain mechanism itself introduce an excess noise which couples to the ROC. This excess noise factor is denoted F.

And lastly, dark currents originating from regions outside the multiplication region[5] also couple to the ROC. This term is denoted I_{DS}, where the subscript S denotes surface leakage currents. Just as the avalanche photocurrent is proportional (by M) to the photocurrent without gain, the amplified noise is also proportional to the noise figure without gain.

In order to derive the SNR, the rms multiplied photocurrent and the shot noise term are written below as shown in [3]. The thermal noise of the ROC is the same as before. The multiplied rms photocurrent is

[5] The multiplication of photogenerated carriers is localized to specific regions of the p-n junction, as shall be shown later.

$$i_P = \frac{q\eta P_0 M}{h\nu\sqrt{2}} \tag{1.55}$$

On the other hand, the shot noise, accounting for the effects of multiplication and excess noise, is given by

$$\left\langle i_S^2 \right\rangle = 2q\left[(I + I_P F_P + I_B F_B + I_D F_D) M^2 + I_{DS}\right]\beta \tag{1.56}$$

The SNR is

$$\frac{S}{N} = \frac{\frac{1}{2}(q\eta P_0/h\nu)^2}{2q(I + I_P F_P + I_B F_B + I_D F_D)\beta + \left[2qI_{DS}\beta/M^2\right] + \left[4kT_{eff}\beta/R_{eq}M^2\right]} \tag{1.57}$$

As Eq. 1.57 shows, the SNR is maximized for large M values, which causes the last two terms of the sum in the denominator to zero. Again, further insight can be gained by solving for the average power as a function of the SNR. This yields

$$P_0 = \frac{2h\nu\beta F_P}{\eta}\left(\frac{S}{N}\right)\left\{1 + \left[1 + \frac{I_{eq}}{q\beta F_P^2 (S/N)}\right]\right\}^{\frac{1}{2}} \tag{1.58}$$

where

$$I_{eq} = I_B F_B + I_D F_D + \frac{I_{DS}}{M^2} + \frac{2kT_{eff}}{qR_{eq}M^2} \tag{1.59}$$

The NEP is found by letting $\frac{S}{N} \rightarrow 1$ and $\beta \rightarrow 1$ in Eq. 1.58:

$$NEP = \frac{2h\nu\beta F_P}{\eta}\left[\frac{I_{eq}}{qF_P^2}\right]^{\frac{1}{2}} \tag{1.60}$$

As Eq. 1.60 shows, large multiplication factors minimize the NEP by eliminating the surface leakage-induced dark current and the thermal noise term. This result implies that increasing the bandwidth does not adversely affect the NEP, unlike in the case of the photodiode. Consequently, APDs are suited for large bandwidth applications [3]. Furthermore, for inherently small R_{eq}, i.e., for high-frequency operation, the internal gain mechanism also minimizes the NEP [3].

1.4.3 Geiger-Mode Avalanche Photodiode Front Ends

In Geiger mode, the APD is biased above the breakdown voltage (Region III in Fig. 1.4). In that mode, the photosensor retains all the same features as the APD that we have just discussed. However, it exploits the discrete nature of photogeneration and as such differs fundamentally from the two previous configurations. It does not measure a current, but rather produces current pulses, the leading edges of which corresponds to the primary carrier that triggered the diverging avalanche current. Its operation is explained below.

A readout circuit and GMAPD is shown in Fig. 1.8. In the case of an APD, it is simply a discrete resistor. At quiescence, the cathode voltage sits at V_{DDA}. If the excess bias voltage (i.e., the voltage differential between the applied voltage and the breakdown voltage) is large enough, a sustainable avalanche current is produced when an electron or a hole is generated within the multiplication region or reaches it by minority carrier diffusion. As a result, the cathode voltage decreases exponentially with a time constant that depends on the reverse bias capacitance and the diode's series resistance R_S. The leading edge of the cathode voltage transient corresponds to the arrival of the avalanche's primary carrier within the multiplication region. For proper Geiger-mode operation, the load resistor R_L is chosen to be large enough so that the avalanche current is quenched within a few hundred picoseconds following the onset of the avalanche process [8].

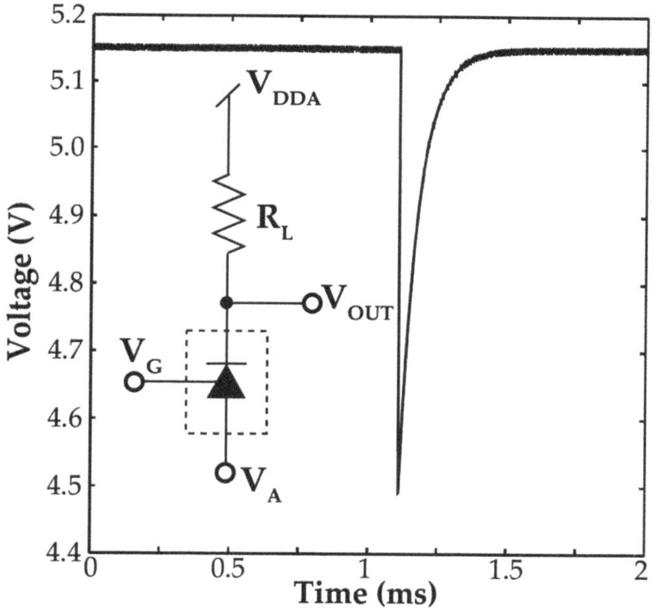

Fig. 1.8 Measured data from a fabricated PGSPAD showing the cathode voltage transient resulting from passive (i.e., a resistor is used as the ROC) Geiger-mode operation for $V_G = -16$ V, $V_A = -13$ V, $V_{DDA} = 5$ V, and $R_L = 100\text{k}\Omega$. (The control voltage V_G will be explained in Chap. 3)

Following this transient decay in the cathode voltage, the diode current is so low that the avalanche process can no longer be sustained; the avalanche current is thus said to be fully quenched and the voltage at which this happens is the quenching threshold. As stated previously, R_L must be large enough to reduce the net reverse bias voltage below the quenching threshold. Failure to do so results in partial quenching and yields increased noise and poor detection efficiency [8].

The quenching threshold is a significant parameter because it implies that the reverse bias voltage need not be reduced below the breakdown voltage in order to quench the avalanche current; this means faster device reset time and consequently faster device operation.

Once the APD is in the quench state, it can no longer detect photons through avalanche and is no longer in Geiger mode [9]. Thus, it must be reset by bringing the cathode voltage to V_{DDA}. In the configuration shown in Fig. 1.8, the leakage current slowly recharges the parasitic capacitance of the APD once it is quenched. This recharge time is comparatively much longer than the quench time because its time constant is dominated by R_L as opposed to much smaller series resistance of the diode (R_S). For fast reset, a transistor can be used to quickly pull the cathode back towards V_{DDA} [8].

1.4.3.1 Geiger-Mode APD Noise

There are several important factors to consider when operating an APD as a SPAD. First, the avalanche process is statistical. This means that the triggering of an avalanche event is associated with a probability measure. It has been shown that this probability increases with increasing excess bias voltage [8, 9]. Second, there are avalanche events that occur in the absence of incident photons. These spurious events constitute the device's internal noise source. They are referred to as dark counts, and the dark count rate (DCR) strongly depends on the number of thermally generated carriers and on the number of carriers trapped from previously occurring avalanches [8, 9]. The latter mechanism is referred to as afterpulsing [8, 9], and in some instances it may be a significant component of the DCR.

1.5 Conclusions

In this chapter we have discussed the mechanisms through which light is converted into an electrical signal. We have reviewed three types of photosensor architectures, namely, the photodiode, the avalanche photodiode, and the Geiger-mode APD. The third configuration allows operation of the photosensor as a photon counter, and for that reason its front end is also called a single-photon avalanche diode (SPAD).

For the three photosensors discussed, the main distinction was the biasing regime. While this is true, a photodiode cannot be operated as an APD or as a SPAD simply by changing the bias regime. There are structural features, for example, that impede proper Geiger-mode operation of p-n junctions that work well as photodiodes. This is especially true in CMOS, where perimeter breakdown confines the multiplication region to the diodes' periphery, which adversely affects their operation in Geiger-mode or as linear APDs. This issue and its prevention is discussed in extensive detail in the chapters below.

References

1. M. Dandin, CMOS Single-Photon Avalanche Diodes and Micromachined Optical Filters for Integrated Fluorescence Sensing, Ph.D. dissertation, University of Maryland, College Park, College Park, Maryland, 2012
2. R. Muller, T. Kamins, M. Chan, *Device Electronics for Integrated Circuits*, 3rd edn. (Wiley, New York, 2003)
3. G.E. Stillman, C.M. Wolfe, *Semiconductors and Semimetals - Infrared Detectors II*, vol. 12 (Academic Press, San Diego, USA, 1977)
4. R.H. Kingston, *Optical Sources, Detectors, and Systems: Fundamentals and Applications* (Academic Press, San Diego, USA, 1995)
5. W.B. Davenport, W.L. Root, *An Introduction to the Theory of Random Signal and Noise* (IEEE Press, New York, 1987)
6. G.M. Williams, A.S. Huntington, *Probabilistic Analysis of Linear Mode vs. Geiger Mode APD FPAs for Advanced LADAR Enabled Interceptors*, ed. by R.T. Howard, R.D. Richards, Proc. SPIE 6220, Spaceborne Sensors III (2006), p. 622008. https://doi.org/10.1117/12.668668
7. A. Van der Ziel, *Noise in Measurements* (Wiley, New York, 1976)
8. A. Gallivanoni, I. Rech, M. Ghioni, Progress in quenching circuits for single photon avalanche diodes. IEEE Trans. Nucl. Sci. **57**(6), 3815–3826 (2010). https://doi.org/10.1109/TNS.2010.2074213
9. S. Cova, M. Ghioni, A. Lotito, I. Rech, F. Zappa, Evolution and prospects for single-photon avalanche diodes and quenching circuits. J. Modern Optics **51**(9–10), 1267–1288 (2004). https://doi.org/10.1080/09500340408235272

Chapter 2
Perimeter-Gated Single-Photon Avalanche Diodes

2.1 Introduction

Avalanche photodiodes consist of p-n junction engineered to withstand high current densities when operated in the near-breakdown regime [1, 2]. APDs biased above breakdown and equipped with suitable readout structures are referred to as single-photon avalanche diodes (SPADs). Due to extremely high electric fields across their depleted junction, SPADs are capable of generating a macroscopic current pulse as a result of a single photogenerated or thermally generated electron-hole pair [3]. With proper noise reduction schemes (cooling or thresholding, for example), they are suitable for photon counting and photon timing applications [4, 5].

SPADs are increasingly being used in biochemical analysis [6–15]. They are favored as an inexpensive optical detection modality due to their ability to transduce weak photon fluxes resulting from fast-decaying optical signals. For instance, they have been used in ultra-weak biochemiluminescence detection [6] and time-resolved fluorescence measurements [8].

One area of research focuses on miniaturizing the hardware used in these bioanalytical techniques in order to create portable devices known as lab-on-a-chip (LOC) [16]. LOCs are microfabricated platforms capable of undertaking complex biochemical sample processing steps and performing a wide range of analyses with little or no interfacing with benchtop equipment. They often comprise microelectromechanical systems (MEMS), and complementary metal oxide semiconductor (CMOS) circuits are favored for use as a means for achieving signal transduction and processing, actuation, and control [17].

There is a strong need to adapt SPAD design techniques to existing CMOS fabrication processes. In addition to widespread commercial availability, CMOS technology offers the possibility to implement a wide variety of functionalities onto a single die. Therefore, CMOS SPADs will lead to enhanced LOC performance by allowing their integration with other on-chip mixed-mode circuits.

M. Dandin et al., *Single-Photon Avalanche Diodes and Photon Counting Systems*,
https://doi.org/10.1007/978-3-031-64334-7_2

Traditionally, SPADs have been fabricated through dedicated processes [18], since achieving well-controlled avalanche profiles in commercial CMOS processes is challenging. As a result of the planar nature of the junctions, electric field distributions show maxima at the diodes' periphery [19]. This means that in the high reverse bias regime, as is needed for SPAD operation, diodes are more susceptible to breakdown at the edges. When the device periphery enters breakdown prematurely, the device never reaches full volumetric breakdown. This adversely results in a reduced active area and, consequently, reduced photon detection efficiency. Therefore, perimeter breakdown is one of the most important obstacles in integrating SPADs into mainstream CMOS processes.

Several researchers have shown successful perimeter breakdown prevention in CMOS avalanche diodes. Rochas et al. have shown, in a twin-well CMOS technology, the use of lateral diffusion of donor atoms following n-well oxidation to effectively create a lighter n-doped region at the edge of the p-n junction [20]. Pauchard et al. have shown a CMOS-compatible process for fabricating a photodiode with reduced perimeter breakdown; the device comprised a field-limiting guard ring placed at a distance d from the p^+ implant in conjunction with a control gate over the gap [21]. We have previously shown the combination of the lateral diffusion of n-wells, similar in spirit to that demonstrated by Rochas et al., in combination with a depletion gate to further reduce premature edge breakdown [22]. All of these techniques stem from original works by Grove et al. and Temple et al. who studied the effects of dopant concentration modulation and junction curvature on the breakdown voltage [19], as well as the effects of field plates over high-field regions [23]. Other architectures in deep-submicron CMOS technologies have made use of shallow trench isolation in order to favorably alter the junction's geometry to prevent edge breakdown [24].

We present herein a thorough analysis of a device which was implemented in a 0.5-μm, single-well, 3-metal, 2-poly, CMOS process. In addition to characterizing the device's breakdown characteristics as a function of gate voltage and lateral diffusion distance, we describe a numerical model which identified regions in which the electron/hole generation rates were maximized, thus indicating locations susceptible to breakdown. This metric (the charge generation rate) allowed us to confirm that the application of the gate voltage, in conjunction with the lateral diffusion of adjacent n-wells, successfully prevented edge breakdown.

We begin with a review of the theory underlying avalanche breakdown in a p-n junction. We then introduce the device architecture and the rationale for the design. Finally, we present a thorough characterization of the device's DC response in the high-reverse bias regime. The empirical results are corroborated by the previously mentioned numerical model.

2.2 Theory of Operation

2.2.1 *Avalanche Breakdown*

When a reverse bias applied across a p-n junction is increased beyond a critical potential, a swift rise in current occurs as a result of a charge multiplication process known as avalanche breakdown. This process is the result of free electrons (holes) attaining kinetic energies high enough in magnitude to ionize lattice atoms upon collisions. Because of the extremely strong electric field present across the junction, the newly generated carriers are in turn accelerated and they, too, reach energies high enough to cause impact ionization. This multiplication cycle continues, giving rise to an avalanche current. The reverse voltage beyond which this phenomenon occurs is the breakdown voltage, and its magnitude is denoted V_B in the remainder of this chapter.

Avalanche breakdown is best described through the theory of impact ionization first postulated by McKay et al. [25, 26]. The number of secondary carriers generated by a moving carrier in an electric field across the space charge region is given by the following integral:

$$1 - \frac{1}{M} = \int_0^w \alpha_i dx \tag{2.1}$$

where M is the multiplication factor that reflects the number of charges collected for each charge generated, α_i is the impact ionization rate (electron [hole] $\times \text{cm}^{-1}$, where the subscript i indicates the carrier type), and w is the width of the space charge region at the applied reverse voltage. Avalanche occurs when, in (1), $M \rightarrow \infty$. Miller subsequently showed the multiplication factor and the device's breakdown voltage to be related as follows:

$$M = \left[1 - \left(\frac{V_A}{V_B}\right)^n\right]^{-1} \tag{2.2}$$

where V_A is the applied voltage and n and V_B are process-dependent constants [27]. As can be seen from Eq. 2.2, $M \rightarrow \infty$ as $V_A \rightarrow V_B$.

Subsequent works have developed a deeper understanding of breakdown mechanisms in semiconductor junctions, and there is a significant body of literature on the topic for a wide range of materials [28–31].

2.2.2 *Low-Light Transduction*

To make use of the avalanche phenomenon and transduce weak photon fluxes with high fidelity using an APD, a suitable readout mechanism must be implemented. One configuration with which this is attained is Geiger-mode operation. In Geiger-mode APDs (GMAPDs), the bias point is well above the junction's breakdown voltage. (As previously mentioned, an APD in such a configuration is a SPAD.) In the absence of free carriers, there is no current, and an extremely high field is present across the junction. Under these conditions, the generation of a free carrier almost immediately triggers an avalanche current through the junction. This current is then quenched using an external circuit, consisting of either a passive ballast resistor or a set of transistors [10]. Quenching brings the bias voltage below the breakdown voltage and effectively stops the avalanche. The junction is then recharged for the detection of another avalanche event. In Geiger mode, the multiplication gain is infinite, and the device functions as a bistable circuit whose transitions are triggered by the generation of a single electron-hole pair. For preliminary work showing Geiger-mode operation of the device described herein, see [22, 32].

2.3 Device Architecture

2.3.1 *Design*

The physical implementation of a SPAD in a standard digital CMOS technology is shown in cross-sectional format in Fig. 2.1 (bottom panel); the top panel illustrates the layout view. Some layers are drawn with half-symmetry for clarity. The avalanche junction is a p^+/n-well diode implanted in p-type bulk substrate material. As the layout indicates, a circular n-well is drawn to serve as the multiplication region. An additional annular n-well structure is drawn at its periphery at a distance denoted by β. Several structures were constructed with values of β varying from 0λ to 7.5λ, in increments of 0.5λ, where $\lambda = 0.35$ μm is half of the feature size of the CMOS technology. We also note that all values of β were smaller than the minimum spacing (8λ) between adjacent n-well regions as prescribed by the design rules.

The anode (p^+ region) is implanted over the multiplication region and extends laterally halfway to the inter-n-well gap. A polysilicon layer serving as a control gate extends from the edge of the anodic implant to partway over the annular n-well structure. A summary of the devices' dimensions is shown in Table 2.1.

2.3.2 *Rationale*

The rationale for this design is twofold; first, with the lateral diffusion of n-wells, a lighter n-doped region is created at the edge of the junction (following

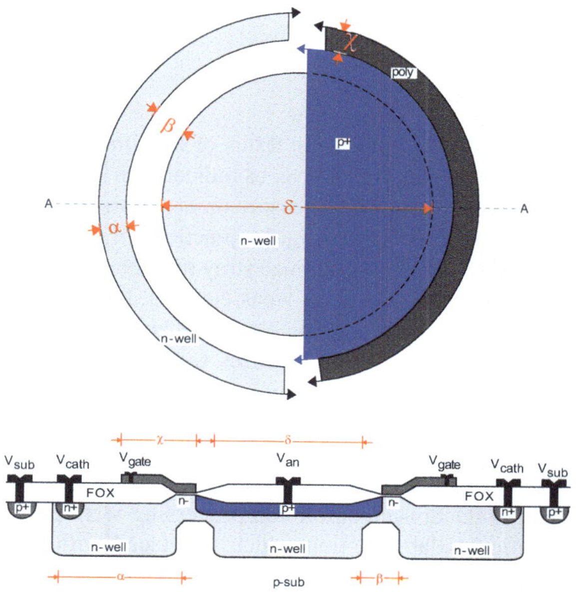

Fig. 2.1 Layout and cross-sectional (along AA) views of the SPAD device. The diode consists of the p^+/n-well junction. A polysilicon gate is placed at the periphery. The adjacent n-wells diffuse into each other following high-temperature steps of the CMOS fabrication process. This effectively forms an n-well region with doping concentration much lower at the periphery than at the center. The gate serves to deplete the corner region of electrons to further reduce the corner electric field (drawing not to scale)

Table 2.1 Device parameters

$\alpha(\mu\text{m})$	$\beta(\lambda)$	$\chi(\mu\text{m})$	$\delta(\mu\text{m})$
9	$0 : 0.5 : 7.5$	3.5	90

p^+implantation). This effectively reduces the corner electric field when compared to the native diodes (devices with $\beta = 0\lambda$ and $V_G = 0$ V). Second, the addition of a gate serves to deplete the corner regions of electrons when a negative bias is applied with respect to the cathode. This creates a field-induced junction which causes the corner electric field to be further reduced.

2.4 Breakdown Characteristics

2.4.1 Experimental Procedures

In this section, we describe the experimental methods that were used for characterizing the devices. We investigated the near-breakdown I-V characteristics of the SPADs as a function of the voltage applied to the control gate. We also sought to understand the effect of the n-well spacing on the I-V characteristics. To do so, devices with β ranging from 0λ to 7.5λ were tested. The voltages used to test the devices are shown in Table 2.2; we use the notation $V_{\text{start}} : V_{\text{increment}} : V_{\text{end}}$ to designate sweep parameters. In the remainder of this chapter, for the sake of simplicity, we shall refer to gate voltages, reverse bias voltages, voltage increments,

Table 2.2 Sweep parameters

V_{an}	V_{cath}	V_{sub}	V_{gate}
$0:-0.03:-18$	0	0	$0:-1:-18$

and breakdown voltages in terms of their magnitudes. Note that the actual polarity of these voltages is negative, as indicated in Table 2.2.

We used two source-measure units (Keithley, 2400), one for biasing the gate and another for reverse-biasing the junction. In the latter case, a four-wire connection was used in order to minimize stray resistances arising from the leads, thus ensuring accuracy when measuring weak currents. For a given device, voltages ranging from 0 to 18 V, in 1-V increments, were applied to the gate, and the I-V curve was collected. (In total, 19 I-V traces were collected.) For each trace, the reverse bias potential across the junction was swept from 0 to 18 V in increments of 30 mV. This voltage increment was kept small in order to accurately resolve the avalanche current. Each measurement from the sweep was collected five times and the average current was recorded (all standard deviations were found to be in the pico-ampere regime). Data acquisition and processing were performed through an application-specific Matlab function which interfaced with Labview subroutines written for instrument control. In order to estimate the breakdown voltage, the Matlab function performed the computations described in further detail in Eqs. 2.3–2.6.

First, the raw current data is transformed into the log domain as $\log(I(V))$ as shown in 2.3. Next, the derivative of the data was computed 2.4. The derivative was normalized 2.5 for display purposes, and the breakdown voltage was taken to be the reverse bias voltage at which the derivative was maximized [2.6]. A moving average may be applied to the data in order to remove any measurement noise, but this was not found to be necessary for any of the data described here. This procedure has also been used for calculating cutoff wavelengths of optical filters [33] and is illustrated graphically in Fig. 2.3.

$$A_n = \log_{10}|I_n|, \quad n = 1, \ldots, N \tag{2.3}$$

$$X_k = \frac{A_{k-1} - A_k}{V_{k-1} - V_k}, \quad k = 2, \ldots, N \tag{2.4}$$

$$\Gamma_i = \frac{X_i}{\max[X]} \tag{2.5}$$

$$V_B = \arg[\max[\Gamma]] \tag{2.6}$$

2.4.2 Results

We collected I-V curves for devices with β ranging from 0λ to 7.5λ; we also varied the gate voltage in order to study its effect.

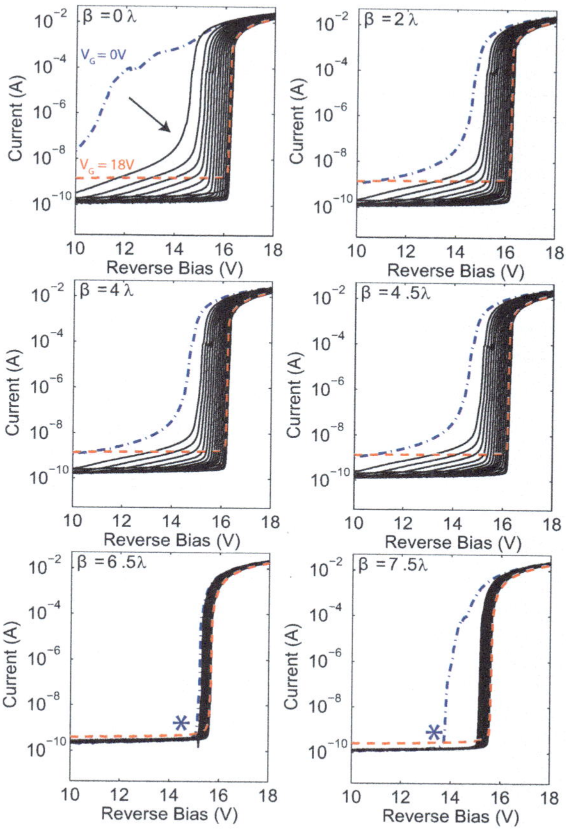

Fig. 2.2 Current-voltage dependence for different devices. These results indicate the effects of both gate and lateral diffusion on the overall device characteristics. In each figure, the black (dash-dotted) line indicates the I-V trace for $V_G = 0$ V and the red (dashed) line indicates the trace for $V_G = 18$ V. The gate voltage is increased by increments of 1 V in all cases in direction of the arrow shown on the first figure (top left). The figures with asterisks indicate cases for devices with $\beta \rightarrow 8\lambda$; at these values of β, the p^+anode tends to short with the substrate which implies that these I-V curves are likely that of the junction formed by the n^+/n-well annulus and the p^+/p-subregions

Holding the gate voltage at 0 V allowed us to determine the effect of n-well lateral diffusion on the avalanche profile. Data for this experiment are indicated in Fig. 2.2 by the dash-dotted lines. As the results show, the native junction, i.e., $\beta = 0\lambda$ and $V_G = 0$ V, exhibits a gradual change in current as the reverse bias is increased. We attribute this excessive leakage to tunneling effects resulting from strong perimeter electric fields. These effects are mitigated as the inter-n-well gap is increased ($\beta = 2, 4$, and 4.5λ in Fig. 2.2). However, at wider gaps ($\beta \rightarrow 8\lambda$), the p^+anode tends to short with the substrate, causing another junction to form between the n^+/n-well annulus and the p^+/p-subregions. This suggests that with the placement of the outer n-well (annulus of width α and inner diameter $\delta + 2\beta$ in Fig. 2.1) alone, one can marginally improve the avalanche profile, as long as the gap is narrow enough to prevent anode to substrate shorting (Fig. 2.3).

Another important implication of Fig. 2.2 is that a large gate potential enhances the devices' rectifying efficiency. This is indicated by the increase in breakdown voltage as the gate voltage is increased (see Fig. 2.4), as well as an increase in the rate of change of the current with reverse voltage. Nevertheless, the gate potential

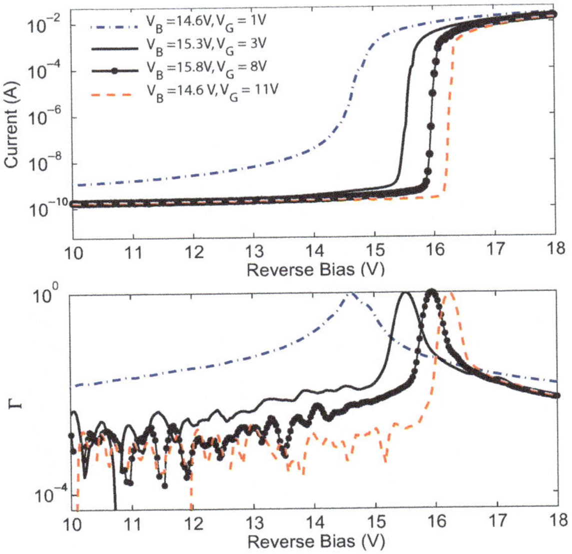

Fig. 2.3 The top panel shows the current as a function of reverse bias. The bottom panel shows the normalized derivative Γ (see Eq. 2.5) for several traces corresponding to the device with $\beta = 0$ for various gate voltages (also shown in Fig. 2.2). In each case, Γ exhibits an extremum. The voltages at which this occurs are identified as the breakdown voltages (see legend in top panel)

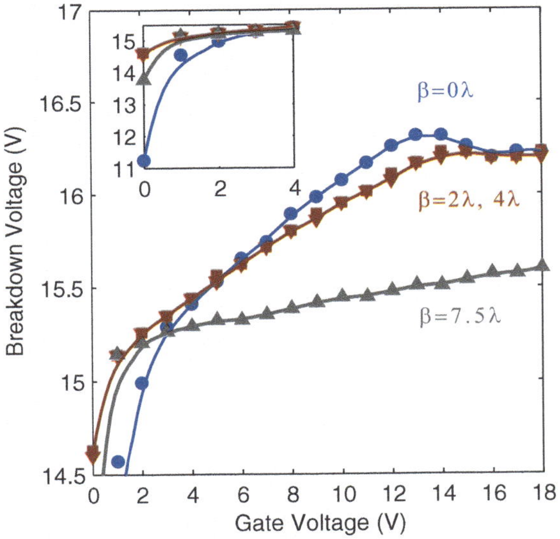

Fig. 2.4 Breakdown voltage as a function of gate voltage for four different structures. The inset indicates a nonlinear relationship for gate voltages ranging from 0 to 4 V. The relationship becomes linear and the breakdown voltage saturates for structures having $\beta = 0, 2, 4\lambda$. However, no saturation is observed for $\beta = 7.5\lambda$, indicating that the lateral diffusion is not as effective and that the two n-wells are more electrically isolated

has a diminished effect for large β, i.e., $(\Delta V_B/\Delta V_G)$ goes down as β is increased. As before, this is due to a conduction path forming from anode to substrate for devices with gaps approaching the technology's recommended minimum separation distance.

Another salient issue is that for gate voltages greater than 17 V, for all devices, there is a significant increase in the pre-breakdown current (red dotted lines). This

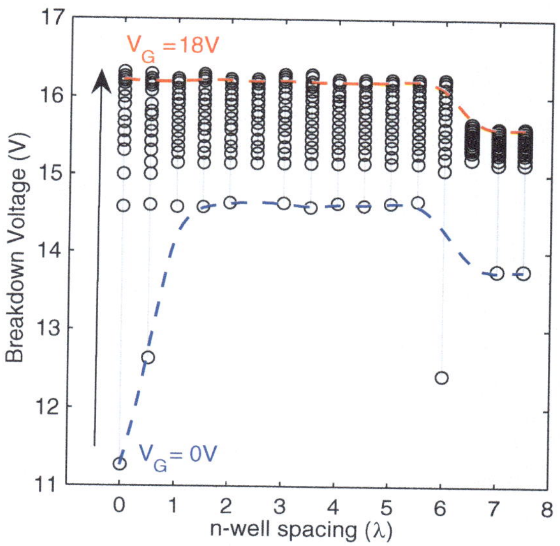

Fig. 2.5 Breakdown voltage as a function of n-well spacing for the eight structures tested. Each circle represents a gate voltage. The gate voltage is increased in magnitude in the direction shown by the arrow in increments of 1 V. The black trace represents averaged data for $V_G = 0$ and the red trace averaged data for $V_G = 18$ V

finding suggests that at these voltages, although perimeter breakdown suppression is achieved, the overall generation rate is increased due to trap-assisted generation. This could lead to increased dark count rate in Geiger-mode operation.

For an overall view of the effect of both suppression techniques, we plotted the breakdown voltage as a function of n-well spacing for each of the gate biases applied in all 16 structures tested. This plot is shown in Fig. 2.5. As the graph indicates, the maximum increase in breakdown voltage is observed in the 0λ case where the $\Delta V_B = 5.3$ V for an 18 V change in the gate potential.

2.5 Avalanche Photodiode Modeling

2.5.1 APD Model

To corroborate the data presented thus far and to identify regions susceptible to avalanche, we developed a drift-diffusion and impact ionization-based two-dimensional device simulator. The simulator self-consistently solves the Poisson equation 2.7 and the electron and hole current continuity equations 2.8 and 2.9 to calculate the electrostatic potential ϕ, the electron concentration n, and the hole concentration p:

$$\nabla^2 \phi = -\frac{q}{\epsilon}(p - n + D) \tag{2.7}$$

$$\frac{\partial n}{\partial t} = \nabla \cdot (-n\mu_n \nabla \phi + \mu_n V_{\text{th}} \nabla n) + G_n \tag{2.8}$$

$$\frac{\partial p}{\partial t} = \nabla \cdot \left(p\mu_p \nabla \phi + \mu_p V_{\text{th}} \nabla p\right) + G_p \tag{2.9}$$

D represents the net ionized dopant density, V_{th} is the thermal voltage, q is the electronic charge, and ϵ is the dielectric constant of the medium. μ and G are the mobility and net generation-recombination rate of a carrier type indicated by the subscript, respectively. The G term incorporates trap-assisted as well as impact-initiated generation-recombination rates.

To characterize the operation and breakdown voltage of the avalanche devices, use of a predictive silicon impact generation rate formula is important. To this end, we employed recently published silicon impact coefficients [34] that are formulated using standard Chynoweth-type expressions with a temperature-dependent exponent 2.10, 2.11. Here, the local electric field E is the field along the total current density (J_t) and T is the absolute temperature. This field is calculated using $|J_t \cdot E_t| / |J_t|$ to remove the part of the local total electric field E_t that does not contribute to the current flow (especially the field normal to the semiconductor-insulator interface):

$$\alpha = n_1 \exp\left[-\left(n_2 + n_3 T\right)/E\right] \tag{2.10}$$

$$\beta = p_1 \exp\left[-\left(p_2 + p_3 T\right)/E\right] \tag{2.11}$$

To calculate electron (hole) impact rate, $n_1\,(p_1)\,, n_2\,(p_2)$, and $n_3\,(p_3)$ are taken as $4.43 \times 10^5 \left(1.13 \times 10^6\right) \text{cm}^{-1}, 9.66\times\ 10^5 \left(1.71 \times 10^6\right) \text{V} \cdot \text{cm}^{-1}$, and $4.99 \times 10^2 \left(1.09 \times 10^3\right) \text{V} \cdot \text{cm}^{-1}\ \text{K}^{-1}$, respectively. These values were obtained empirically using silicon p-i-n diodes [32]. Furthermore, the band-to-band electron (hole)-initiated impact rate used in the current continuity equations is calculated using the electron (hole) current density $J_n\left(J_p\right)$ in $G_n = \alpha J_n/q \left(G_p = \beta J_p/q\right)$.

2.6 Doping Profile Model

Breakdown voltages, dark current levels, and multiplication efficiencies are all significantly affected by the doping profiles of the n-well and the p^+-n^+ terminals. As the doping on either side of a p-n junction increases, the breakdown voltage decreases due to higher associated built-in electric fields. The externally applied reverse bias results in a field that adds to this built-in field and therefore gives rise to increases in carrier kinetic energies and eventually impact ionizations.

To be able to predict performance of our APDs, we first need to estimate doping profiles of the wells and the substrate in the technology that our devices are fabricated in. As these doping details are proprietary information, we resort to

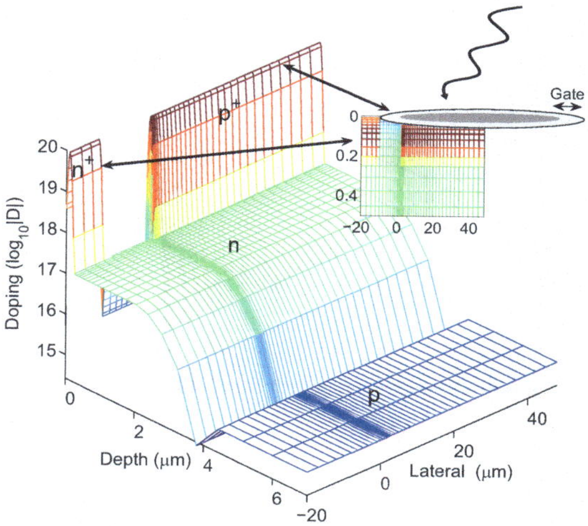

Fig. 2.6 Cross section of half of the doping profile of the perimeter-gated APD with adjacent *n*-wells (0λ case). Inset shows the location of the gate and an incoming photon, combining bird's-eye and side views

empirical and numerical methods that give reasonable matches between calculated and measured data. Additionally, the extraction of a two-dimensional doping profile using experimental methods such as C-V measurements is often problematic as they provide a one-dimensional average doping map at a depletion region edge that is traced over the entire device by changing biases. In Fig. 2.6, we show the inferred doping map of an APD with zero spacing between its two n-wells ($\beta = 0\lambda$ in Fig. 2.1). This doping profile gives a good match between experimental and measured data and is in line with doping profiles deduced for this process using other device configurations [35]. More specifically, the peak concentration in $p^+$and n^+ regions is 1×10^{20} cm^{-3}, and this concentration drops with $\sigma = 50$ nm and $\sigma = 120$ nm in the depth and lateral directions assuming a shifted Gaussian profile where the mean is 0.1 μm away from the surface. For the n-well, the empirically obtained maximum doping is 1.22×10^{17} cm^{-3}, and it also has a shifted Gaussian profile in the depth direction. These doping profiles result in a p-n junction at approximately 0.2-μm depth for the p^+terminal in an n-well. Here, the substrate is set to 1×10^{15} cm^{-3}, and this gives rise to a doping crossover at roughly 3.5-μm depth for the n-well surrounded by the p substrate.

2.7 Simulation and Model Validation

To investigate the effects of the perimeter gate and doping profile on breakdown characteristics of our APD, we simulate a diode with the doping profile shown in Fig. 2.6. The simulated current-voltage curves along with the generation rate profile at 17-V reverse bias (i.e., −17 V is applied to the p^+anode, and the $n^+$and substrate nodes are grounded) and 0-V gate bias are shown in Fig. 2.7a. Here, the

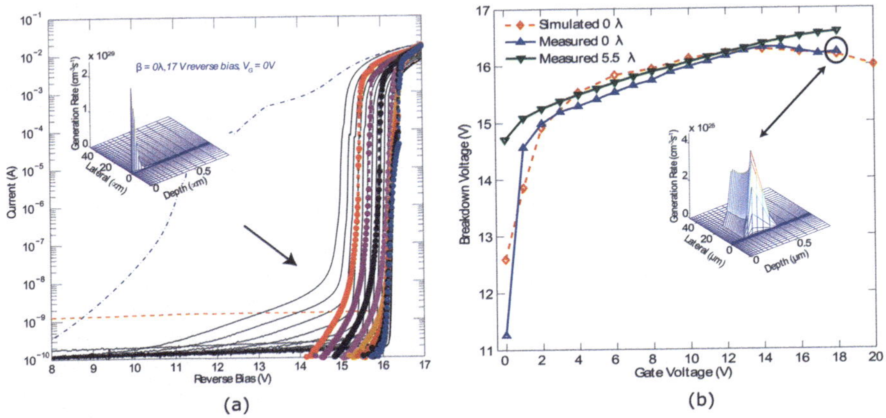

Fig. 2.7 (**a**) Current-voltage characteristics of an APD with a zero n-well spacing (0λ). The doping profile used in simulations is shown in Fig. 2.6. The inset depicts the generation profile at 0-V gate voltage and 17-V reverse bias. For 0-V gate bias, the generation rate peaks near the semiconductor-insulator interface in the depth direction and the p+-n edge in the lateral direction. (**b**) Room temperature breakdown voltage as a function of gate bias for the APD with zero n-well spacing, as well as another APD with a 5.5λ n-well spacing. We compare experimental results with those of a simulated 0λ device. APDs with other well spacings show similar characteristics. Also, the inset shows that at gate biases higher in magnitude, breakdown moves away from the surface and occurs at the p+-n corner as well as laterally along the p+-n edge away from the surface

gate bias is changed by 2-V increments from 0 V to 18 V. Comparison of measured and simulated currents shows a reasonable match. For the APDs designed and tested and also simulated, the spacing between the n-wells changes the breakdown voltage modestly (unless the n-wells are far apart and the channel under the gate shorts the p^+ tap to the substrate). However, the gate voltage noticeably affects the breakdown voltage, as well as the dark current level and the sharpness of the current change during breakdown. Thus, we performed simulations for the 0λ design under different gate and reverse biases and temperatures.

As the reverse bias is changed from 0 to 10 V, overall generation rate starts rising near the surface, as the lowest resistance path between $p^+$and n^+nodes runs parallel to the interface for low applied biases. However, the current as well as the generation rate is low for this bias range. For the 0-V gate bias case, as the reverse bias is gradually increased, the field and generation rate near the silicon-silicon dioxide interface close to the p^+-n edge start increasing, and this eventually results in avalanche breakdown near the surface, as shown in Fig. 2.7a (inset). Figure 2.7b shows experimental and simulated breakdown voltages as functions of gate biases. The breakdown voltage is calculated as the point where the curvature of the current-voltage curve changes sign, as previously described in (3)–(6). The comparison shows a good match between simulated results for the 0λ device and measured data for the 0λ and 5.5λ devices. For the gate voltages shown in Fig. 2.7a and b that are higher in magnitude, as the reverse bias voltage increases, the lateral field at the surface decreases because the voltage bias between the gate and the p^+anode is reduced.

This causes current to spread outward through the edges, as well as through the corner of the p^+anode. Thus, the highest electric field region is near the p^+anode and n-well corner, since the corner shows a combination of both the lateral and vertical change in local electric field. When we further increase the reverse voltage, the corner electric field as well as the field in the normal direction of the p^+-n edge away from the interface rises rapidly. The generation rate at these locations due to current flow and high electric field gives rise to secondary carriers that are high enough in numbers to affect the terminal currents. A positive feedback between the current and multiplication rate is balanced by the slight reduction in electric field at the corner and along the deeper edge thus resulting in a finite current at the output.

However, when there are abundant impact-generated carriers, the junction fields and the device's rectification efficiency drop significantly, giving rise to a resistive current-voltage behavior at very high reverse biases. For the high aforementioned gate biases, APDs break down at the p^+-n corner and along the p^+-n edge deeper into the substrate, as shown in Fig. 2.7b (inset). The breakdown characteristics at gate biases higher in magnitude are in line with our initial design principles, which are verified by these simulations.

For a given negative gate bias, holes are attracted under the gate. We note that for gate biases lower than -18 V, the channel under the gate has a large hole concentration. For a wide spacing between the n-wells, this channel results in an undesirable short between the p^+region and the substrate. Also, when gate biases greater than 18 V are applied to devices with overlapping n-wells, breakdown voltages start decreasing as the lowering of the lateral field at the surface and the vertical field becomes less effective in pushing the current flow and breakdown region down into the substrate.

We further show in Fig. 2.8 the simulated and measured gate bias breakdown voltages for a 5.5λ device at 50°C and 75°C. The breakdown voltage increases with rising temperature, since the built-in electric fields as well as mobilities slightly decrease with an increase in intrinsic carrier concentration and phonon population at high temperatures. Figure 2.8 (inset) shows that the silicon electron and hole impact rates that are calculated using 2.10 and 2.11 decrease with increasing temperature. This is consistent with high scattering rates which occur at higher temperatures.

2.8 Perimeter-Gated SPAD Modeling

While perimeter gating has been shown to effectively curtail PEB and provide high-SNR Geiger-mode operation [36], a unified equivalent circuit model is needed to incorporate its use in circuit and system designs. The following sections describe our early efforts towards developing such models. First, we provide a survey of key SPAD modeling approaches, highlighting important model features. Second,

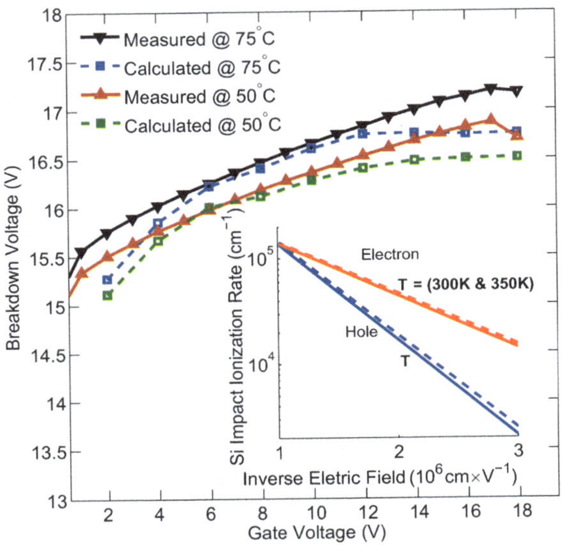

Fig. 2.8 Experimental and simulated breakdown voltages as a function of gate bias at 50° and 75°C. We used a 5.5λ device for measurements. Inset shows that the silicon impact ionization coefficient decreases with rising temperature for both types of carriers

we describe a perimeter-gated SPAD model and its use in designing a Geiger-mode front end fabricated in 0.5-μm standard CMOS process.

2.8.1 *State-of-the-Art SPAD Modeling Approaches*

Existing SPAD models can be grouped in four classes, which are labeled (1) through (4) hereinafter. Each class is associated with a specific level of the design cycle. The models are shown in Table 2.3, and they are tabulated according to their respective class and key features.

Class 1 models simulate SPADs at the semiconductor level, making use of semiclassical or purely quantum-mechanical formulations of the transport properties of the semiconductor structure underlying the SPAD. Class 1 models yield key operational parameters such as breakdown voltage, avalanche gain (useful for APD operation), multiplication noise, etc. For example, in our own work, we have made use of a class 1 model to determine the effects of structural features on the breakdown voltage and the localization of the avalanche region within a p+/n-well junction [1].

Class 2 models are equivalent circuit representations of the SPAD, incorporating passive circuit components, and, in some instances, behavioral voltage sources and/or behavioral current sources. Class 2 models reproduce both DC I-V characteristics and Geiger-mode temporal characteristics. They are used primarily for circuit-level designs and simulation. Furthermore, class 2 models allow the simulation of readout circuits based on the SPAD's electrical characteristics (internal resistance,

Table 2.3 State-of-the-art SPAD models

References	Model class	Model features
Mita et al. [37]	2	Verilog-A
Arbat et al. [38]	3	Equivalent SPAD circuit
Giustolisi et al. [39]	3	Verilog-A
Gonzalez et al. [40]	4	Equivalent SiPM circuit
Garcia et al. [41]	3	Verilog-A
Therrien et al. [42]	4	Python and Matlab
Pancheri et al. [43]	1	Dead space and impact ionization
Chick et al. [44]	3	Dead time compensation
Dandin et al. [1, 36]	1	Perimeter-gate effects on breakdown voltage
Hurkx et al. [45]	1	Tunneling
Zappa et al. [46]	2	SPICE
Kolb [47]	3	Probabilistic model, Monte Carlo simulations
Mora et al. [48]	2	PSPICE SPAD model
Javitt et al. [49]	1	Breakdown state stability
Spinelli et al. [50]	1	Physics-based numerical analysis
Kang et al. [51]	1	Physical analytical model
Wang et al. [52]	1	Breakdown probabilities
Zhao et al. [53]	1	Negative feedback analytical model
Lacaita et al. [54]	1	Monte Carlo simulation of avalanche
Tan et al. [55]	1	Random ionization path length
Masudy-Panah et al. [56]	1	Analytical model of velocity enhancement
Vanyushin et al. [57]	1	Discrete analytical model
Corsi et al. [58]	4	Equivalent circuit of SiPM
Zaghloul et al. [59]	1	TCAD device modeling
Dolgos et al. [60]	1	Full-band Monte Carlo, PIN devices
McFarlane [61]	4	Information theoretic approach
Field et al. [62]	4	Probabilistic figure of merit
Ramirez et al. [63]	4	Information theoretic approach
This work	2	pg-SPAD equivalent circuit

anode-to-substrate capacitance, cathode-to-substrate capacitance, and reverse bias capacitance).

As such, class 2 models readily provide a first-order evaluation of performance at the circuit level. For example, with class 2 models, parameters such as readout latency, quenching time constant, and reset time constant can be estimated.

Class 3 models are also used for circuit-level designs, but they differ from class 2 models in that they account for the stochastic nature of Geiger-mode operation. As such, they provide a more realistic simulation platform. Class 3 models make use of hardware description language that rely on theoretical (or empirically derived) formulations of dark count rate, breakdown probability, signal-to-noise ratio (with or without dead time correction or temporal gating), noise-equivalent power, and after-pulsing probability.

Class 4 models are used in system-level designs. These models often incorporate a combination of the previously mentioned models to simulate array behavior and arrive at system-level metrics. These system-level models are useful for simulating the operation of multiple SPADs, whether they are configured in low-density or high-density arrays.

The class 4 models can also be used to study operational aspects of SPAD arrays, such as pixel cross-talk, array breakdown voltage nonuniformity, etc. In rare instances, system-level metrics can also be studied at the single-pixel level. Examples include models that consider the SPAD front end as an information channel from which information-theoretic metrics like information capacity and mutual information can be derived as a function of operational parameters like excess bias voltage, dark count rate, and afterpulsing probability [61, 63].

The model featured in the present work is a class 2 pg-SPAD model. As shown in Table 2.3, previous pg-SPAD models were either device physics models or system-level models. The class 2 pg-SPAD model provides the needed simulation infrastructure to support the development of sensor arrays and readout circuits, as shown in the case study developed in Section III. The model's key device parameters were obtained experimentally. These parameters can alternatively be obtained from a class 1 model of the pg-SPAD, like the one reported in [1]. In the latter case, the model is hierarchical because it inherently includes class 1 model parameters.

Before describing the class 2 pg-SPAD model in detail, the following section discusses the theory of operation of pg-SPAD in DC and in Geiger mode from the perspective of its operation in a front-end circuit. For simplicity, a passive quenching and reset readout circuit is described.

2.8.2 Perimeter-Gated SPAD Theory of Operation

Figure 2.9 illustrates the DC and transient characteristics of a perimeter-gated avalanche photodiode (APD) that was discussed in great length in [1]. (For clarity, it is noted that when the perimeter-gated APD is operated in Geiger mode, it is referred to as a pg-SPAD.) Figure 2.9a shows the I-V profile of the perimeter-gated APD for gate voltages ranging from 0 to 18 V in magnitude, in 1-V increments. (In total, 19 I-V traces were measured.) For each trace, the reverse bias potential across the junction was swept from 0 to 18 V in increments of 30 mV. This voltage increment was kept small in order to accurately resolve the avalanche current. Each measurement from the sweep was collected five times and the average current was recorded (all standard deviations were found to be in the pico-ampere regime) [1].

These data show that the perimeter-gated APD is effectively an APD with *variable* breakdown voltage. While this tunability may be desired for some applications, the device is useful as a SPAD only in an *effective range* of perimeter-gated voltage, as shall be discussed below.

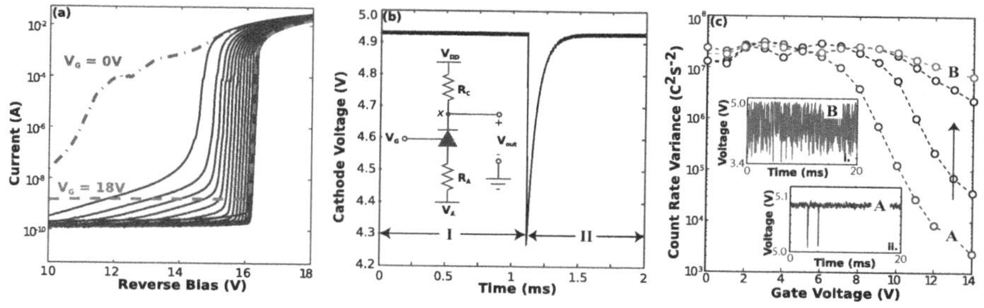

Fig. 2.9 (**a**) DC characteristics of a 35-μm-diameter pg-SPAD for different gate voltages [1] (reused with permission). (**b**) Transient behavior of a pg-SPAD. The circuit configuration is shown in the inset. In this configuration, $V_{DD} = 5$ V, $V_A = -13$ V, $R_C = 100$ kΩ, $R_A = 50$ Ω, and $V_G = -13$ V. Regions **I** and **II** indicate the quiescent and recharge phase, respectively. The pg-SPAD is said to be quenched between regions **I** and **II**. (**c**) Dark count rate variance as a function of gate voltage. Inset **i** shows dark pulses in a 10-ms time frame for trace A at $V_G = -13$ V, for an excess bias of $V_E = 0.5$ V, whereas inset **ii** shows dark pulses in a 10 ms for trace B at $V_G = -13$ V, for an excess bias of $V_E = 1.5$ V. The two traces between trace A and trace B correspond to traces for 0.5 V $< V_E <$ 1.5 V, with the arrow indicating increasing V_E

Unlike conventional APD pixels, SPAD pixels make use of the transient behavior of the APD. That is, the APD must be biased in Geiger mode. An example Geiger-mode configuration is shown in the inset of Fig. 2.9b. In that regime of operation, the junction is reverse-biased with a voltage of magnitude greater than its breakdown voltage V_B.

At quiescence (region **I** in Fig. 2.9b), the cathode (node x) voltage is at a voltage substantially equal to V_{DD}. When a free carrier is generated in the depleted junction, that carrier undergoes multiplication through impact ionization avalanche mechanisms. The voltage at node x decreases exponentially with a time constant that depends on the effective capacitance at the cathode and the diode series resistance. The leading edge of the transient voltage at node x corresponds, within tens to hundreds of picoseconds, to the arrival of the free carrier in the depleted junction.

Following the current increase and the cathode voltage drop, the voltage across the APD becomes small because of the ohmic voltage drop developed across the resistor R_C, which is large enough that avalanche can no longer occur. In this phase, the APD is said to be quenched. It then enters the reset phase (region **II** in Fig. 2.9b). In that phase, capacitive displacement currents slowly recharge the effective capacitance at node x. This recharge phase is exponential in time, and its time constant is dominated by R_C, which means that resetting a quenched APD using the configuration shown in Fig. 2.9 takes much longer than quenching it. After reset, the APD is again at the quiescent level and ready to register the arrival of another free carrier.

In practical applications, the number of pulses like the one shown in Fig. 2.9b can be counted during a predetermined time period, and the count (minus the average count due to noise) is correlated with the number of photons incident on the SPAD's active area during that time period. The count rate, i.e., the number of pulses per unit time, corresponds to the photon arrival rate scaled by an efficiency factor that is less than 1. This efficiency factor is governed by the detector's photon detection efficiency, spectral responsivity, quantum efficiency, and other structural factors.

An important performance factor, however, is the detector's dark count, which is a noise process originating from free carriers whose emissions are not triggered by photons. Specifically, these free carriers may be generated as a result of thermal generation processes or as a result of the release of trapped carriers that are temporally correlated with avalanche events preceding their release. These processes constitute the primary mechanisms contributing to the noise of the SPAD, and thus, they are the main causes of the dark count rate.

Previous publications have shown that perimeter gating decreases the dark count [1]. However, as shown in Fig. 2.9c, the variance in the dark count rate starts to decrease, only after applying 6 V on the gate. In other words, the gate voltage must be sufficiently large to effectively reduce the dark count rate using perimeter gating. As such, in the perimeter-gated SPAD model described in this chapter, the effective perimeter-gate voltage range is taken to be from $V_G = 6$ V to $V_G = 17$ V in magnitude.

2.8.3 Perimeter-Gated SPAD Equivalent Circuit Model

Figure 2.10a shows an equivalent circuit model typically used for simulating an APD in Geiger mode [64]. In this model, R_D is the series resistance formed by the resistance of the diode's depleted junction and that of its neutral regions, V_B is the diode's breakdown voltage, C_{AC} is the reverse bias junction capacitance, C_{AS} is the anode to substrate capacitance, and C_{CS} is the cathode to substrate capacitance.

In this model, the avalanche process is modeled by closing switch S_A and by subsequently opening switch S_A, which corresponds to ending the avalanche process. As was pointed out by Mora et al. [46, 48], this model does not accurately capture the diode's temporal dynamics in the Geiger mode because the avalanche duration depends on manually setting the time during which switch S_A stays closed. Stated otherwise, the model of Fig. 2.10a does not capture the self-sustaining and self-quenching nature of the avalanche process.

The avalanche is self-quenching when the junction current is low enough that carriers crossing the multiplication region are too few to cause impact ionization. In that situation, an avalanche current cannot be maintained through the diode. On the other hand, if the current is high enough, the avalanche is sustained until it is quenched externally. As noted in [64], a Geiger-mode APD self-quenches if the current is less than $I_q = 100\ \mu$A; otherwise, the current is sustained, and it must be quenched using passive or active methods.

An improved Geiger-mode model that captures the self-quenching and self-sustaining features of the avalanche process is shown in Fig. 2.10b [46]. In this model, switch S_A is modeled using two ideal voltage-dependent switches (S_1 and S_{TRIG}) and one current-dependent switch (S_{SELF}).

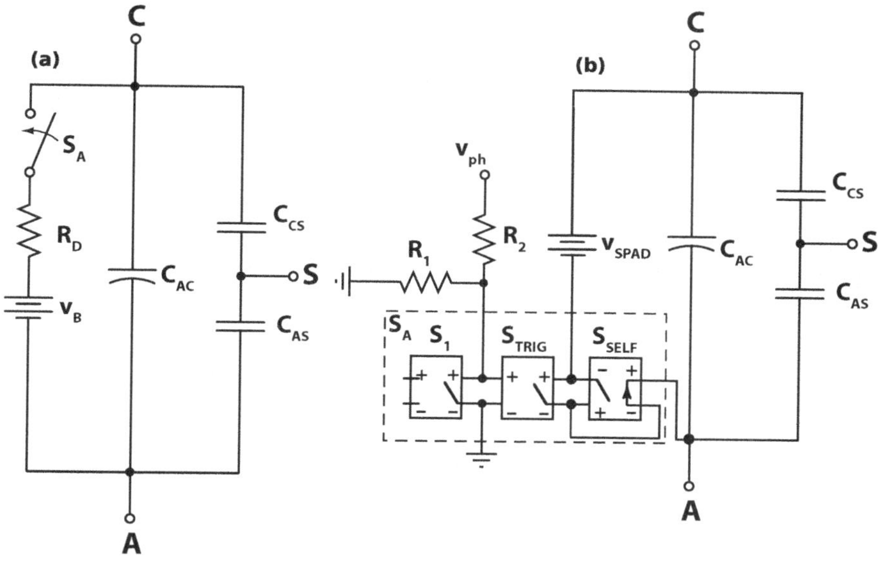

Fig. 2.10 (**a**) Equivalent circuit model of a generic Geiger-mode APD [64]. (**b**) Improved Geiger-mode APD equivalent circuit model with self-sustaining and self-quenching capability [46]

In the improved model, the current-controlled switch S_{SELF} has its threshold set at I_q. The threshold of switch S_1 is set at $-V_B$ in order to prevent avalanche triggering, should the device's reverse voltage fall below the breakdown voltage. The improved model further accounts for the reverse bias dependence of the diode resistance R_D.

The reverse bias dependence of R_D is modeled using nonlinear voltage generator V_{SPAD}, which approximates the breakdown curve using a piecewise linear relationship between the current and the voltage; this is implemented with a piecewise linear voltage source with parameters consisting of points selected along the breakdown curve. Lastly, the photon stimulus is modeled either as a pulsed voltage source v_{ph} or as a pulsed current source i_{ph} (not shown in Fig. 2.10). A voltage divider (R1 and R2) ensures that any one of the two sources can be used, without modification to the model. In this improved model, S_{TRIG} serves to trigger the avalanche current. As such, its duration is no longer dependent on the duration of the stimulus but rather on the level of the avalanche current flowing through S_{SELF}.

While the model of Fig. 2.10b accurately captures the Geiger-mode temporal dynamics, it cannot be used to simulate a perimeter-gated structure. For example, the improved model cannot account for the change in breakdown voltage resulting from changing the gate voltage observed in a perimeter-gated SPAD, as illustrated in Fig. 2.9a and discussed above and in [1, 22, 36].

An equivalent circuit model that is adequate for a perimeter-gated SPAD is shown in Fig. 2.11. While based on the improved SPAD equivalent model of Zappa et al.

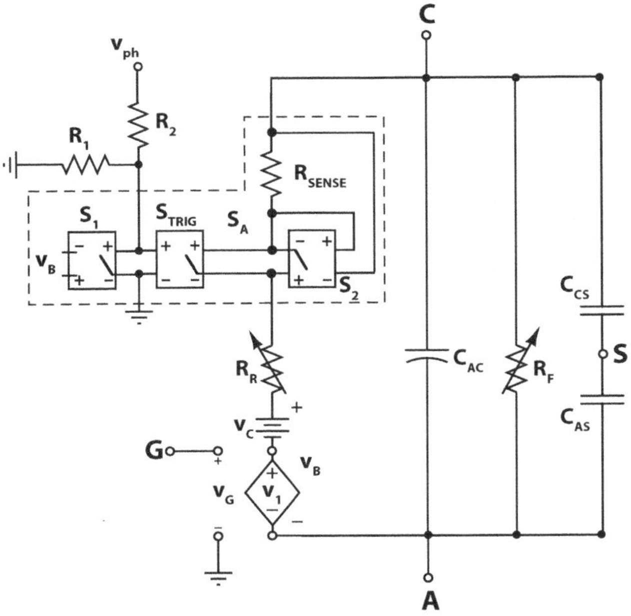

Fig. 2.11 Perimeter-gated single-photon avalanche diode equivalent circuit model. The terminals A, C, S, and G are the anode, cathode, substrate, and gate terminals, respectively

Table 2.4 pg-SPAD model parameters

Symbol	pg-SPAD1	pg-SPAD2
A	7×10^{11}	4×10^{12}
B	−8.461	−18.16
λ	−12.15	−13.60
C	7×10^{11}	4×10^{12}
D	−22.7	−27.18
C_{CS}	0.5–1 pF	0.5–1 pF
C_{AS}	2–3 pF	2–3 pF
C_{AC}	15–25 pF	15–25 pF

[46], this model includes the gate voltage dependence and the variable reverse bias resistance that are only particular to a pg-SPAD.

The model is based on equations that are shown in Eqs. 2.12, 2.13, and 2.14, which can easily be incorporated in a SPICE netlist to provide circuit-level simulation capabilities. The model parameters for two different pg-SPADS are reported in Table 2.4, and the model results, showing good agreement with measured DC and Geiger-mode responses, are shown for the two pg-SPADs in Fig. 2.12.

In the perimeter-gated SPAD model, the avalanche is simulated the same way as in Fig. 2.10b, but the pg-SPAD equivalent circuit includes components that are used to model behaviors specific to a perimeter-gated SPAD. Namely, the reverse

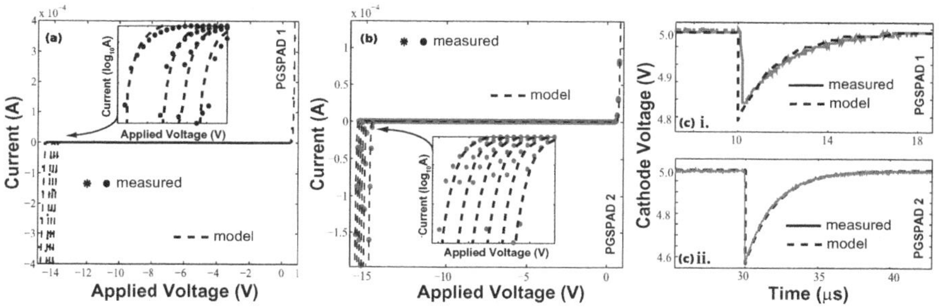

Fig. 2.12 (**a**) Measured DC characteristics of pg-SPAD1 for different gate voltages and superimposed with the pg-SPAD model results. The inset shows a close-up of the near breakdown region, showing good agreement between the model and the measured data. (**b**) Measured DC characteristics of pg-SPAD2 for different gate voltages and superimposed with the pg-SPAD model results. The inset shows a close-up of the near breakdown region, showing good agreement between the model and the measured data. (**c**) Measured Geiger-mode responses with passive quenching and reset readout, superimposed with the Geiger-mode response obtained from the model, for both devices

bias resistance is modeled as a variable resistance that depends on the reverse bias voltage and on the gate voltage.

The perimeter-gated SPAD breakdown voltage is modeled as a voltage-dependent source whose control terminal is the gate terminal of the perimeter-gated SPAD. The functional relationship between the breakdown voltage V_B and the gate voltage V_G may be derived experimentally by plotting the breakdown voltage versus the gate voltage and by using a suitable method for calculating V_B from the measured I-V characteristics [1].

For example, based on the measured data obtained for the two different pg-SPADs reported in Table 2.4, the breakdown voltage was found to depend on the zero-bias breakdown voltage, V_{BD_0}, and a voltage $V_1 = \gamma \times V_G$, where γ is a fitting constant obtained from the perimeter-gate voltage versus breakdown voltage curve (Fig. 2.4 in [1]). In the model, V_{AC} and V_{CA}, they reflect the anode-to-cathode voltage and the cathode-to-anode voltage, respectively:

$$V_B = V_1 + V_{BD_0} \tag{2.12}$$

$$R_R = Ae^{B(V_{CA}-V_{BD_0}+V_1)} \tag{2.13}$$

$$R_F = Ce^{DV_{AC}} \tag{2.14}$$

2.9 Geiger-Mode Readout

The perimeter-gated SPAD model discussed above may be used to design and simulate pixels and pixel arrays in CMOS processes. In the remainder of the chapter, we report on the design of a CMOS SPAD pixel coupled with an on-chip digital readout circuit. The described pixel was designed and fabricated in a standard 3-metal, 3-poly, single-well CMOS process.

Figure 2.13a illustrates a front end that includes a perimeter-gated SPAD and its associated readout circuit, and Fig. 2.13b shows the same circuit in greater detail. In Fig. 2.13a, the dashed line surrounds the components that form the active quenching and reset circuit (AQAR). The AQAR module includes a pFET MR that is used for resetting the pg-SPAD. Specifically, when MR is turned on, the voltage at node x is pulled towards V_{DD}, effectively creating a large applied voltage approximately equal to $(V_{DD} - V_{LOW})$ across the pg-SPAD. For Geiger-mode operation the applied voltage must exceed the pg-SPAD's breakdown voltage. Thus, to reset the pg-SPAD, the signal R is asserted by a reset logic circuit (RLC) to turn on MR at a time coinciding with the end of the quenching phase. Further, the time duration of the reset (t_{RESET}) may be adjusted using the RLC.

Controlling the reset duration is necessary because MR must remain on for a period of time that is equal to or exceeds the RC time constant required to pull up node x to V_{DD}. If MR is turned off too early, the pg-SPAD will not be fully reset, and it may thus operate with an efficiency that is different than the efficiency observed in the previous cycle. The controllable reset duration feature also allows for compatibility of the readout circuit with pg-SPADs of different sizes. Such flexibility is required because pg-SPADs of different sizes have different effective capacitances at node x, thereby requiring different reset and quenching times.

The pg-SPAD is quenched by asserting the signal Q, turning on the nFET MQ, which effectively pulls down node x to the quenching voltage V_Q. The quenching voltage is chosen so that the applied voltage across the pg-SPAD $(V_Q - V_{LOW})$

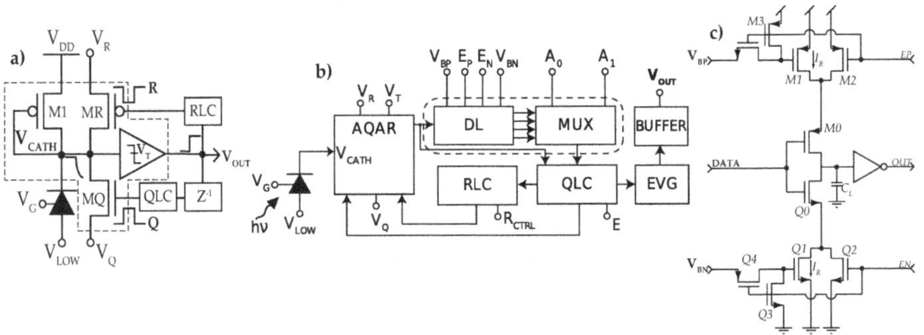

Fig. 2.13 (**a**) The active quenching and active readout (AQAR) circuit includes transistors M1, MR, and MQ. The transistors MR and MQ serve to reset and quench the pg-SPAD, respectively. (**b**) Pixel block diagram. The readout circuit (ROC) includes an active quenching and active reset module (AQAR) and a delay generator (dashed line) that includes a voltage-controlled delay line and a 4-to-1 multiplexer. The ROC also includes a reset logic circuit (RLC), a quenching logic circuit (QLC), an event generator (EVG), and a digital readout buffer. (**c**) Variable delay element including a top delay regulator (M1–M4) and a bottom delay regulator (Q1–Q4), which provide two delay regimes, based on V_{BP}, V_{BN}, EP, and EN

is small enough that an avalanche cannot occur. It is noted that while quenching is mostly achieved using an active component, namely, MQ, the initial quenching phase is passive, and it is provided by the resistance of the diode-connected pFET M1.

The quenching phase may be held for a predetermined amount of time, i.e., for the duration of a hold-off time denoted $t_{HOLDOFF}$. Holding the PSGSPAD in the quenched state ensures that carriers trapped during the avalanche do not trigger additional avalanche events. These carriers are not related to photon arrival in the junction but are rather a by-product of the avalanche process; this phenomenon is termed afterpulsing. Therefore, holding the junction in the quenched state for a time $t_{HOLDOFF}$ that exceeds the trapped carrier lifetime contributes to reducing the dark counts that are due to afterpulsing.

In order to generate $t_{HOLDOFF}$, a quenching logic circuit (QLC) is used. The QLC may be programmed to hold the voltage at node x at V_Q for a time period equal to $t_{HOLDOFF}$. The hold-off time may be generated using a monostable circuit, or as in the present implementation, with a delay module triggered by the leading edge of a digital pulse produced by a Schmitt trigger coupled to node x. Once the voltage at node x falls below a predetermined threshold V_T of the Schmitt trigger, the quenching signal Q is asserted and the delay module (denoted Z^{-1} in Fig. 2.13a) is engaged.

The delay module includes a series of variable delay elements (VDEs). The circuit schematic of the VDE is shown in Fig. 2.13c. Each VDE produces a fixed delay amount t_{VDE}, which is bounded by a minimum achievable delay, t_{MIN}, and a maximum achievable delay t_{MAX}. The delay module may include n VDEs disposed in series, thereby providing the capability to produce a maximum hold-off time equal to $n \times t_{VDE}$. An n-to-1 multiplexer may be used to select a particular output node of the series, which enables hold-off times equal to $(n - k) \times t_{VDE}$ to be produced, wherein $k = 1, \ldots, (n - 1)$.

In the present implementation, a 4-stage delay module (i.e., n = 4) is used, and each one of the VDEs is designed to have $t_{MIN} = 70$ ns and $t_{MAX} = 700$ ns. Further, in the present work, the analog control voltages of the current-starved inverter (denoted V_{BP} and V_{BN}) may be chosen so as to produced a coarse delay variation regime and a fine delay variation regime (see Fig. 2.13c).

In the coarse regime, the delay of each VDE, and consequently that of the whole delay module, may be changed with tens of nanoseconds of resolution. In the fine delay variation regime, the delay time may be varied using nanosecond resolution. The present architecture allows $t_{HOLDOFF}$ to be varied from 70 ns to 2. 8 μs, with one of the two aforementioned programmable resolutions. One additional feature allows the QLC to operate with $t_{HOLDOFF} = 0$, effectively bypassing the delay module. In this mode, the pixel is said to be free-running, and it may exhibit relatively higher dark count rates. Nevertheless, with strong enough signals, operating in free-running mode may yield higher count rates and thus an improved dynamic range.

The operation of the AQAR circuit is controlled by the reset voltage V_R, the quenching voltage V_Q, and the Schmitt trigger threshold voltage V_T. The delay line

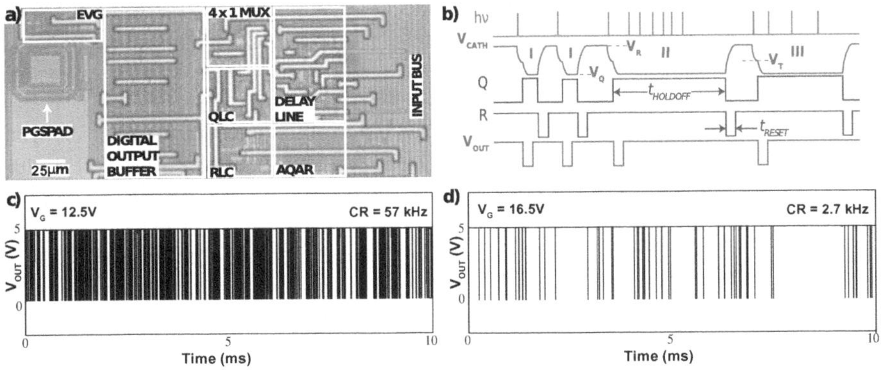

Fig. 2.14 (**a**) Photomicrograph of the pixel. The perimeter-gated SPAD is monolithically integrated with the readout circuit (ROC). (**b**) Timing diagram illustrating the pixel's operation. (**c**, **d**) Measured results from the fabricated pixel, in the dark

is indicated by the dashed line in Fig. 2.13b (the 4 VDEs are not shown). The digital control bits A0 and A1 allow the 4-to-1 multiplexer to select different stages of the delay line. The analog control voltages V_{BP} and V_{BN} control the amount of the delay produced in each cell, where the delay amount is bounded as explained above. Moreover, the digital control bits EP and EN allow the selection of the delay variation regime, i.e., of the selection of either a coarse delay variation resolution or that of a fine delay variation resolution. The control bit E is used to toggle the QLC in hold-off mode and in free-running mode, and R_{CTRL} is an analog voltage used to control the duration of t_{RESET}.

Figure 2.14a shows a photomicrograph of the pixel. Further, Fig. 2.14b shows a typical timing diagram of the pixel during operation. The first trace is representative of a random photon signal impinging on the PGPSPAD's active area. This photon signal may be modeled using a pulsed voltage source or a pulsed current source when simulating the front end using the pg-SPAD model of Fig. 2.11. The trace V_{CATH} indicates the voltage at node x, for three different hold-off times (denoted I, II, and III) that may be generated by the delay module. The traces Q and R represent the quenching and reset signals which control the gates of MQ and MR, respectively. The V_{OUT} trace represents the digital output signal corresponding to the photons that are registered by the device. As can be seen from Fig. 2.14b, photons arriving during a hold-off period are not registered at the output.

Further, it is noted that the output voltage is produced using an event generator (EVG) block that produces a digital pulse whose leading edge corresponds in time with the time at which the quenching signal Q is asserted, i.e., when the voltage of node x crosses the V_T threshold. The duration of the pulses in the V_{OUT} trace has no significance; only their leading edges is important since these edges correspond to the onset of avalanche. This design is particularly useful when the pg-SPAD behavior is to be buffered and observed off-chip. In many applications, a counter may be used to count the pulses of the pg-SPAD, and a digital count may be read out at the end of a predetermined monitoring period.

Figure 2.14c and d shows measured dark count rate results from the pixel for two different gate voltages. As shown in these figures, a gate voltage of 12.5 V magnitude yields a count rate of 57 kHz. However, once the gate voltage magnitude is increased to 16.5 V, the dark count rate is reduced to 2.7 kHz, which corresponds to a 21-fold reduction in the dark count rate. Additional reduction in dark count rate may be achieved by further reducing the device size and/or by cooling the substrate.

2.10 Conclusions

In this chapter, we reported on the breakdown characteristics of a single-photon avalanche diode structure fabricated in a 0.5-μm single-well CMOS process. Our investigation focused on two methods used for preventing edge breakdown in planar junctions; they were the placement of a field ring at the perimeter to create field-induced junctions and force the maximum electric field to be located at the middle of the junction, and the lateral diffusion of adjacent n-wells, through the deliberate violation of the design rules in the fabrication process. We first conducted a phenomenological study which revealed increases in breakdown voltages resulting from both methods independently.

Following this investigation, we showed that the charge generation rate was a suitable metric for identifying whether perimeter breakdown was prevented. We demonstrated its use through a self-consistently solved 2-D numerical model based on Poisson's equation and the hole and electron current continuity equations coupled with rate equations for carrier generation due to impact ionization. The model showed that the generation rate was maximized (a condition indicating susceptibility to avalanche) at the corner, confirming that in the native diodes, the lines of maximum electric fields are located at the corners. The model also confirmed that breakdown suppression techniques employed were adequate for curtailing edge breakdown in junctions fabricated in this process. Particularly, the field ring at the perimeter proved to be the most effective at preventing perimeter breakdown. This was evidenced by a spatial broadening (away from the surface and into the substrate) of the charge generation function which implies that volumetric breakdown was achieved.

Furthermore, we presented an equivalent circuit model for perimeter-gated single-photon avalanche diodes. The model was developed from measured characteristics of fabricated pg-SPADs, and it was shown to accurately simulate the DC and Geiger-mode behaviors of the fabricated devices. Furthermore, the chapter presented the design of a pixel that includes a pg-SPAD and a readout circuit specifically designed for characterizing the pixel. The readout circuit has the capability of producing variable hold-off times with two programmable resolutions, thereby allowing a detailed characterization of afterpulsing characteristics.

We also provided an overview of current SPAD modeling approaches. The survey revealed that most models are class 1 models, focusing on SPAD device physics rather than on circuit simulations. This trend may be indicative of ongoing efforts to fabricate SPADs in a wide variety of processes, a task that understandably requires the development of device physics models. Interestingly, several of the class 1 models of Table 2.3 focus on device physics aspects that go beyond the transport equations. One can infer from this that there remains a number of research avenues to be explored at the device physics level.

Further, Table 2.3 indicates that despite the many classes of models that exist, there is no *unified* SPAD model. In other words, while some models may be partly hierarchical, there is no single compact model that can account for all levels of the design cycle. This stems from the wide variety of SPAD architectures that exist, in addition to the wide variety of applications in which SPADs are used, all of which require different constraints.

A unified model would obviate the need for a bottom-up approach in design by providing the ability to account, at any given time, for the effects of parameters associated with a specific design level on overall performance. It would be desirable for future models to provide hierarchical descriptions of SPAD operation so that they might be used for optimization from device to system-level designs.

References

1. M. Dandin, A. Akturk, B. Nouri, N. Goldsman, P. Abshire, Characterization of single-photon avalanche diodes in a 0.5 μm standard CMOS process—Part 1: perimeter breakdown suppression. IEEE Sensors J. **10**(11), 1682–1690 (2010). https://doi.org/10.1109/JSEN.2010.2046163
2. M. Dandin, M.H.U. Habib, B. Nouri, P. Abshire, N. McFarlane, Characterization of single-photon avalanche diodes in a 0.5 μm standard CMOS process—Part 2: Equivalent circuit model and Geiger mode readout. IEEE Sensors J. **16**(9), 3075–3083 (2016). https://doi.org/10.1109/JSEN.2016.2526665
3. S. Cova, M. Ghioni, A. Lotito, I. Rech, F. Zappa, Evolution and prospects for single-photon avalanche diodes and quenching circuits. J. Modern Opt. **51**(9–10), 1267–1288 (2004). https://doi.org/10.1080/09500340408235272
4. C. Niclass, A. Rochas, P.-A. Besse, E. Charbon, Toward a 3-D camera based on single photon avalanche diodes. IEEE J. Sel. Topics Quantum Electron. **10**(4), 796–802 (2004). https://doi.org/10.1109/JSTQE.2004.833886
5. C. Niclass, A. Rochas, P.A. Besse, E. Charbon, Design and characterization of a CMOS 3-D image sensor based on single photon avalanche diodes. IEEE J. Solid State Circuits **40**(9), 1847–1854 (2005). https://doi.org/10.1109/JSSC.2005.848173
6. M. Gersbach, Y. Maruyama, C. Niclass, K. Sawada, E. Charbon, A Room Temperature CMOS Single Photon Sensor for Chemiluminescence Detection, in *ConferenceInternational Conference on Miniaturized Systems for Chemistry and Life Sciences (MicroTas)*, Tokyo, Japan, 11 2006, pp. 1–2
7. E. Charbon, Towards large scale CMOS single-photon detector arrays for lab-on-chip applications. J. Phys. D Appl. Phys. **41**(9), 094010 (2008). https://doi.org/10.1088/0022-3727/41/9/094010

8. B.R. Rae, C. Griffin, J. McKendry, J.M. Girkin, H.X. Zhang, E. Gu, D. Renshaw, E. Charbon, M.D. Dawson, R.K. Henderson, CMOS driven micro-pixel LEDs integrated with single photon avalanche diodes for time resolved fluorescence measurements. J. Phys. D Appl. Phys. **41**(9), 094011 (2008). https://doi.org/10.1088/0022-3727/41/9/094011
9. E. Grigoriev, A. Akindinov, M. Breitenmoser, S. Buono, E. Charbon, C. Niclass, I. Desforges, R. Rocca, Silicon photomultipliers and their bio-medical applications. Nucl. Instrum. Methods Phys. Res. Sect. A Accelerat. Spectrom. Detect. Assoc. Equip. **571**(1–2), 130–133 (2007). https://doi.org/10.1016/j.nima.2006.10.046
10. H. Finkelstein, M.J. Hsu, S. Zlatanovic, S. Esener, Performance trade-offs in single-photon avalanche diode miniaturization. Rev. Sci. Instrum. **78**(10), 103103 (2007). https://doi.org/10.1063/1.2796146
11. I. Rech, S. Cova, A. Restelli, M. Ghioni, M. Chiari, M. Cretich, Microchips and single-photon avalanche diodes for DNA separation with high sensitivity. ELECTROPHORESIS **27**(19), 3797–3804 (2006). https://doi.org/10.1002/elps.200600144
12. I. Rech, A. Restelli, S. Cova, M. Ghioni, M. Chiari, M. Cretich, Microelectronic photosensors for genetic diagnostic microsystems. Sens. Actuators B Chem. **100**(1–2), 158–162 (2004). https://doi.org/10.1016/j.snb.2003.12.073
13. I. Rech, G. Luo, M. Ghioni, H. Yang, X. Xie, S. Cova, Photon-timing detector module for single-molecule spectroscopy with 60-ps resolution. IEEE J. Sel. Topics Quantum Electron. **10**(4), 788–795 (2004). https://doi.org/10.1109/JSTQE.2004.833975
14. B. Ozmen, E.U. Akkaya, Infrared fluorescence sensing of submicromolar calcium: pushing the limits of photoinduced electron transfer. Tetrahedron Lett. **41**(47), 9185–9188 (2000). https://doi.org/10.1016/S0040-4039(00)01662-2
15. G. Hungerford, D.J.S. Birch, Single-photon timing detectors for fluorescence lifetime spectroscopy. Meas. Sci. Technol. **7**(2), 121–135 (1996). https://doi.org/10.1088/0957-0233/7/2/002
16. A. Manz, N. Graber, H. Widmer, Miniaturized total chemical analysis systems: A novel concept for chemical sensing. Sens. Actuators B Chem. **1**(1–6), 244–248 (1990). https://doi.org/10.1016/0925-4005(90)80209-I
17. H. Lee, D. Dam, R.M. Westervelt, *CMOS/Microfluidic Hybrid Systems* (Springer, New York, 2007)
18. B.F. Aull, A.H. Loomis, D.J. Young, R.M. Heinrichs, B.J. Felton, P.J. Daniels, D.J. Landers, Geiger-mode avalanche photodiodes for three-dimensional imaging. Lincoln Lab. J. **13**(2), 335–350 (2002)
19. V. Temple, M. Adler, Calculation of the diffusion curvature related avalanche breakdown in high-voltage planar p-n junctions. IEEE Trans. Electron Dev. **22**(10), 910–916 (1975). https://doi.org/10.1109/T-ED.1975.18241
20. A. Rochas, A. Pauchard, P.-A. Besse, D. Pantic, Z. Prijic, R. Popovic, Low-noise silicon avalanche photodiodes fabricated in conventional CMOS technologies. IEEE Trans. Electron Dev. **49**(3), 387–394 (2002). https://doi.org/10.1109/16.987107
21. A. Pauchard, P.-A. Besse, R.S. Popovic, Simulation of a New CMOS Compatible Method to Enhance the Breakdown Voltage of Highly Doped Shallow pn Junctions, in *Proc. Int. Conf. Modeling and Simulation of Microsystems*, 1998, pp. 420–425
22. M. Dandin, N. Nelson, V. Saveliev, H. Ji, P. Abshire, I. Weinberg, Single Photon Avalanche Detectors in Standard CMOS, in *2007 IEEE Sensors* (IEEE, 2007), pp. 585–588. https://doi.org/10.1109/ICSENS.2007.4388466. ISBN:978-1-4244-1261-7
23. A.S. Grove, O. Leistiko, W.W. Hooper, Effect of surface fields on the breakdown voltage of planar silicon p-n junctions. IEEE Trans. Electron Dev. **14**, 157–162 (1967)
24. H. Finkelstein, M.J. Hsu, S.C. Esener, STI-bounded single-photon avalanche diode in a deep-submicrometer CMOS technology. IEEE Electron Dev. Lett. **27**(11), 887–889 (2006). https://doi.org/10.1109/LED.2006.883560
25. K.G. McKay, Avalanche breakdown in silicon. Phys. Rev. **94**(4), 877–884 (1954). https://doi.org/10.1103/PhysRev.94.877

26. K.G. McKay, K.B. McAfee, Electron multiplication in silicon and Germanium. Phys. Rev. **91**(5), 1079–1084 (1953). https://doi.org/10.1103/PhysRev.91.1079
27. S.L. Miller, Ionization rates for holes and electrons in silicon. Phys. Rev. **105**(4), 1246–1249 (1957). https://doi.org/10.1103/PhysRev.105.1246
28. J.Y. Tang, K. Hess, Impact ionization of electrons in silicon (steady state). J. Appl. Phys. **54**(9), 5139–5144 (1983). https://doi.org/10.1063/1.332737
29. A.G. Chynoweth, Ionization rates for electrons and holes in silicon. Phys. Rev. **109**(5), 1537–1540 (1958). https://doi.org/10.1103/PhysRev.109.1537
30. R. Kokosa, R. Davies, Avalanche breakdown of diffused silicon p-n junctions. IEEE Trans. Electron Dev. **ED-13**(12), 874–881 (1966). https://doi.org/10.1109/TED.1966.15862
31. S.L. Miller, Avalanche breakdown in Germanium. Phys. Rev. **99**(4), 1234–1241 (1955). https://doi.org/10.1103/PhysRev.99.1234
32. B. Nouri, M. Dandin, P. Abshire, Characterization of Single-Photon Avalanche Diodes in Standard CMOS, in *Proceedings of IEEE Sensors* (2009). https://doi.org/10.1109/ICSENS.2009.5398384
33. M. Dandin, P. Abshire, E. Smela, Optical filtering technologies for integrated fluorescence sensors. Lab Chip **7**(8), 955 (2007). https://doi.org/10.1039/b704008c
34. D. Massey, J. David, G. Rees, Temperature dependence of impact ionization in submicrometer silicon devices. IEEE Trans. Electron Dev. **53**(9), 2328–2334 (2006). https://doi.org/10.1109/TED.2006.881010
35. A. Akturk, J. Allnutt, Z. Dilli, N. Goldsman, M. Peckerar, Device modeling at cryogenic temperatures: Effects of incomplete ionization. IEEE Trans. Electron Dev. **54**(11), 2984–2990 (2007). https://doi.org/10.1109/TED.2007.906966
36. M. Dandin, P. Abshire, High signal-to-noise ratio avalanche photodiodes with perimeter field gate and active readout. IEEE Electron Dev. Lett. **33**(4), 570–572 (2012). https://doi.org/10.1109/LED.2012.2186112
37. R. Mita, G. Palumbo, P. Fallica, Accurate model for single-photon avalanche diodes. IET Circuits Dev Syst. **2**(2), 207 (2008). https://doi.org/10.1049/iet-cds:20070180
38. A. Arbat, A. Dieguez, D. Gascon, J. Trenado, L. Garrido, Avalanche Photodiodes for High Energy Particle Tracking in 130 nm CMOS Technology, in *2009 16th IEEE International Conference on Electronics, Circuits and Systems - (ICECS 2009)* (IEEE, 2009), pp. 579–582. https://doi.org/10.1109/ICECS.2009.5410849. ISBN:978-1-4244-5090-9
39. G. Giustolisi, R. Mita, G. Palumbo, Behavioral modeling of statistical phenomena of single-photon avalanche diodes. Int. J. Circuit Theory Appl. **40**(7), 661–679 (2012). https://doi.org/10.1002/cta.748
40. A.J. Gonzalez, M. Moreno, J. Barbera, P. Conde, L. Hernandez, L. Moliner, J.M. Monzo, A. Orero, A. Peiro, R. Polo, M.J. Rodriguez-Alvarez, A. Ros, F. Sanchez, A. Soriano, L.F. Vidal, J.M. Benlloch, Simulation study of resistor networks applied to an array of 256 SiPMs. IEEE Trans. Nucl. Sci. **60**(2), 592–598 (2013). https://doi.org/10.1109/TNS.2012.2226051
41. M. Moreno Garcia, O.s. Guerra Vinuesa, R. del Rio Fernandez, B. Perez Verdu, A. Rodriguez Vazquez, CMOS SPADs Selection, Modeling and Characterization Towards Image Sensors Implementation, in *2012 19th IEEE International Conference on Electronics, Circuits, and Systems (ICECS 2012)* (IEEE, 2012), pp. 332–335. https://doi.org/10.1109/ICECS.2012.6463734. ISBN:978-1-4673-1260-8
42. A.C. Therrien, B.-L. Berube, C. Thibaudeau, S. Charlebois, R. Lecomte, R. Fontaine, J.-F. Pratte, Modeling of Single Photon Avalanche Diode Array Detectors for PET Applications, in *2011 IEEE Nuclear Science Symposium Conference Record* (IEEE, 2011), pp. 48–53. https://doi.org/10.1109/NSSMIC.2011.6154399. ISBN:978-1-4673-0120-6
43. L. Pancheri, D. Stoppa, G.-F. Dalla Betta, Characterization and modeling of breakdown probability in sub-micrometer CMOS SPADs. IEEE J. Sel. Topics Quantum Electron. **20**(6), 328–335 (2014). https://doi.org/10.1109/NSSMIC.2011.6154399
44. S. Chick, R. Coath, R. Sellahewa, R. Turchetta, T. Leitner, A. Fenigstein, Dead time compensation in CMOS single photon avalanche diodes with active quenching and external

reset. IEEE Trans. Electron Dev. **61**(8), 2725–2731 (2014). https://doi.org/10.1109/TED.2014.2332068
45. G. Hurkx, H. de Graaff, W. Kloosterman, M. Knuvers, A new analytical diode model including tunneling and avalanche breakdown. IEEE Trans. Electron Dev. **39**(9), 2090–2098 (1992). https://doi.org/10.1109/16.155882
46. F. Zappa, A. Tosi, A.D. Mora, S. Tisa, SPICE modeling of single photon avalanche diodes. Sens. Actuators A Phys. **153**(2), 197–204 (2009). https://doi.org/10.1016/j.sna.2009.05.007
47. K. Kolb, Signal-to-noise ratio of Geiger-mode avalanche photodiode single-photon counting detectors. Opt. Eng. **53**(8), 081904 (2014). https://doi.org/10.1117/1.OE.53.8.081904
48. A. Dalla Mora, A. Tosi, S. Tisa, F. Zappa, Single-photon avalanche diode model for circuit simulations. IEEE Photon. Technol. Lett. **19**(23), 1922–1924 (2007). https://doi.org/10.1109/LPT.2007.908768
49. M. Javitt, V. Savuskan, T. Merhav, Y. Nemirovsky, Revisiting single photon avalanche diode current-voltage modeling and transient characteristics. J. Appl. Phys. **115**(20), 204503 (2014). https://doi.org/10.1063/1.4875855
50. A. Spinelli, A. Lacaita, Physics and numerical simulation of single photon avalanche diodes. IEEE Trans. Electron Dev. **44**(11), 1931–1943 (1997). https://doi.org/10.1109/16.641363
51. Y. Kang, H.X. Lu, Y.-H. Lo, D.S. Bethune, W.P. Risk, Dark count probability and quantum efficiency of avalanche photodiodes for single-photon detection. Appl. Phys. Lett. **83**(14), 2955–2957 (2003). https://doi.org/10.1063/1.1616666
52. S. Wang, F. Ma, X. Li, G. Karve, X. Zheng, J.C. Campbell, Analysis of breakdown probabilities in avalanche photodiodes using a history-dependent analytical model. Appl. Phys. Lett. **82**(12), 1971–1973 (2003). https://doi.org/10.1063/1.1559946
53. K. Zhao, S. You, J. Cheng, Y.-h. Lo, Self-quenching and self-recovering InGaAsInAlAs single photon avalanche detector. Appl. Phys. Lett. **93**(15), 153504–153504-3 (2008). https://doi.org/10.1063/1.3000610
54. A. Lacaita, S. Cova, A. Spinelli, F. Zappa, Photon-assisted avalanche spreading in reach-through photodiodes. Appl. Phys. Lett. **62**(6), 606–608 (1993). https://doi.org/10.1063/1.108870
55. C.H. Tan, J.S. Ng, G.J. Rees, J.P.R. David, Statistics of avalanche current buildup time in single-photon avalanche diodes. IEEE J. Sel. Topics Quantum Electron. **13**(4), 906–910 (2007). https://doi.org/10.1109/JSTQE.2007.903843
56. S. Masudy-Panah, V.A. Tikkiwal, Velocity enhancement in APDs with sub-100-nm multiplication region. Opt. Commun. **346**, 167–171 (2015). https://doi.org/10.1016/j.optcom.2015.02.043
57. I.V. Vanyushin, V.A. Gergel, V.M. Gontar', V.A. Zimoglyad, Y.I. Tishin, V.A. Kholodnov, I.M. Shcheleva, A discrete model of the development and relaxation of a local microbreakdown in silicon avalanche photodiodes operating in the Geiger mode. Semiconductors **41**(6), 718–722 (2007). https://doi.org/10.1134/S1063782607060206
58. F. Corsi, A. Dragone, C. Marzocca, A. Del Guerra, P. Delizia, N. Dinu, C. Piemonte, M. Boscardin, G. Dalla Betta, Modelling a silicon photomultiplier (SiPM) as a signal source for optimum front-end design. Nucl. Instrum. Methods Phys. Res. Sect. A Accelerat. Spectrom. Detect. Assoc. Equip. **572**(1), 416–418 (2007). https://doi.org/10.1016/j.nima.2006.10.219
59. B. Zhang, Z. Li, M.E. Zaghloul, A Single-Photon Avalanche Diode in CMOS 0.5μm n-Well Process, in *2012 IEEE Sensors* (IEEE, 2012), pp. 1–4. https://doi.org/10.1109/ICSENS.2012.6411333. ISBN:978-1-4577-1767-3
60. D. Dolgos, H. Meier, A. Schenk, B. Witzigmann, Full-Band Monte Carlo Simulation of Single Photon Avalanche Diodes, in *2013 IEEE Photonics Conference* (IEEE, 2013), pp. 360–361. https://doi.org/10.1109/IPCon.2013.6656586. ISBN:978-1-4577-1507-5
61. N. McFarlane, An Analysis of the Information Efficiency of Single Photon Avalanche Diodes, in *2012 IEEE 55th International Midwest Symposium on Circuits and Systems (MWSCAS)* (IEEE, 2012), pp. 730–733. https://doi.org/10.1109/MWSCAS.2012.6292124. ISBN:978-1-4673-2527-1

62. R.M. Field, J. Lary, J. Cohn, L. Paninski, K.L. Shepard, A low-noise, single-photon avalanche diode in standard 0.13μm complementary metal-oxide-semiconductor process. Appl. Phys. Lett. **97**(21), 211111 (2010). https://doi.org/10.1063/1.3518473
63. D. Ramirez, M. Hayat, S. Torres, B. Saleh, M. Teich, Information-theoretic criterion for the performance of single-photon avalanche photodiodes. IEEE Photon. Technol. Lett. **17**(10), 2164–2166 (2005). https://doi.org/10.1109/LPT.2005.856406
64. S. Cova, M. Ghioni, A. Lacaita, C. Samori, F. Zappa, Avalanche photodiodes and quenching circuits for single-photon detection. Appl. Opt. **35**(12), 1956 (1996). https://doi.org/10.1364/AO.35.001956

Chapter 3
Optoelectronic Characteristics of Perimeter-Gated Single-Photon Avalanche Diodes

3.1 Introduction

As noted in the previous chapter, when biased in the near-breakdown regime, CMOS p-n junctions suffer from premature edge breakdown (PEB) [4, 5], limiting their usefulness as Geiger-mode avalanche photodiodes (APDs). PEB suppression mechanisms have been widely studied in the literature [6–9]. Typically, PEB is averted by implanting a guard ring on the periphery of the photodiode in order to ensure breakdown at the volumetric junction instead of on the peripheral junction. Unfortunately, in small devices, the guard ring occupies a large portion of the photodiode's area, which results in a significant reduction in pixel fill factor. Alternative PEB suppression methods which do not adversely affect fill factor have consisted of using shallow trench isolation (STI) to physically taper the periphery of the diode [8] or placing the avalanche junction in a retrograde buried n-well implant in order to widen the depletion regions at the boundary of the APD's anode [9].

Chapter 2 reported two methods for suppression of perimeter breakdown for a 90-μm-diameter APD fabricated in a three-metal two-poly 0.5-μm single-well mixed-signal CMOS process [10]. The device featured two mechanisms for reducing peripheral electric fields: a laterally diffused n-well guard ring and a polysilicon field gate over the perimeter of the diode. We found that the field gate proved most effective at reducing perimeter breakdown and, subsequently, that the technique scaled easily, with no fill-factor loss, to a $25 \times 25\ \mu m^2$ APD.

Perimeter gating also provides a means for actively and dynamically changing the p-n junction's breakdown voltage. Manipulating the breakdown voltage is specifically important in the case of single-photon avalanche diodes (SPADs). SPADs are p-n junctions biased in Geiger mode, where a single charge carrier can start an avalanche current upon reaching the depletion layer [11]. PEB causes an increase in the SPAD's dark count rate (DCR) and reduced photon detection efficiency [12, 13]. Perimeter gating enables tuning breakdown voltages of the SPADs to yield improved DCR performance, offset fabrication non-idealities from

M. Dandin et al., *Single-Photon Avalanche Diodes and Photon Counting Systems*,
https://doi.org/10.1007/978-3-031-64334-7_3

pixel to pixel, and counter breakdown voltage shifts that occur during operation as a result of local or array-wide temperature variations.

Perimeter gating affects the SPAD's spectral responsivity as well. The relationship between junction type, field effects, and spectral responsivity is well known. For example, Merill developed color sensors based on multi-junction front ends in which different junction depths are used to extract color information without the aid of color filters [14]. Moreover, field gates and transistor gates are also well known as means for controlling photocurrent gain and spectral response [15–17]. Similarly, perimeter gating influences spectral responsivity since it is a field effect that readily modulates local carrier concentrations, generation rates, and junction widths [1, 10, 18].

Here, we feature several contributions relating to the optoelectronic characteristics of pg-SPADs and pg-SPAD-based devices: (1) we demonstrate the monolithic interfacing of active quenching and reset circuits with a perimeter-gated single-photon avalanche diode (pg-SPAD) similar to the structure reported in [10]; (2) we demonstrate a corollary effect of PEB suppression, namely, more than one order of magnitude reduction in DCR which allows Geiger-mode operation with a high signal-to-noise ratio (SNR); (3) we show that a large negative bias on the perimeter gate allows the reliable detection of a continuous monochromatic beam of low intensity $\left(16\ \mathrm{nW/cm^2}\right)$ with an SNR of three, whereas, under high-illumination conditions $\left(5\mu\mathrm{W/cm^2}\right)$, the detector's SNR exceeds 300, which is 37 times higher than when a bias voltage 6 V lower in magnitude is applied to the perimeter gate; (4) we estimate the spectral responsivity of a pg-SPAD as a function of applied junction voltage and gate voltage and show that perimeter gating strongly influences the spectral responsivity by broadening the response over the visible range; and (5) we report on the spectral responsivity and on the photoresponse nonuniformity (PRNU) of a 64 × 64 pg-SPAD imager.

3.2 Device Architecture

3.2.1 Front-End and Readout Circuit Design

The 0.5-μm test chip comprised 16 channels operated in parallel. Each channel produced a digital pulse train, the number of pulses corresponding to the activity of the detector. Figure 3.1 shows the architecture and operation of a single channel which comprised a pg-SPAD, a thresholding circuit with a threshold set at V_T, active reset and active quenching transistors (MR and MQ), reset and quenching logic circuits (RLC and QLC), and an event generation (EVG) circuit for producing a digital pulse whose leading edge corresponds to an avalanche event.

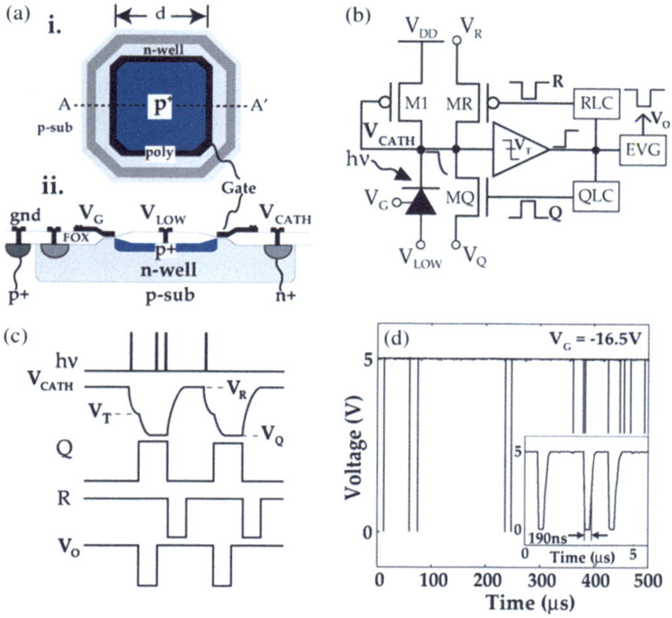

Fig. 3.1 (**a**) Structure of the p + /n-well pg-SPAD [10] (top view shown in *i* and cross-sectional view along AA' in ii). $d = 25$ μm. (**b**) Circuit architecture. (**c**) Timing diagram ($V_R = 5$ V, $V_Q = 2.5$ V, and $V_T = 4.0$ V). (**d**) Sample output trace of a single channel in the dark (the inset shows the same data over a shorter time period)

3.2.2 Readout Circuit Operation

At quiescence, the cathode voltage [V_{CATH} in Fig. 3.1b is near V_R; V_{LOW} is set at -13 V which results in an 18 V reverse bias across the pg-SPAD, corresponding to an excess bias voltage of $V_E = 1.15$ V. Moreover, the gates of the reset and quenching transistors MR and MQ are tied to the outputs of the RLC and the QLC, respectively. These two circuits are monostable in the OFF state of each transistor, i.e., the output of the QLC sits at 0 V and that of the RLC sits at 5 V, in the absence of an avalanche event.

When an event occurs, V_{CATH} drops exponentially towards the ground. When it crosses the V_T threshold, the quenching signal is asserted. MQ is turned on, and the cathode is pulled down to V_Q which is chosen so that the reverse bias of the diode in the quench state is below the junction's breakdown voltage. The cathode remains at V_Q for a fixed time period after which the reset signal is asserted and MR pulls the cathode to V_R. The channel dead time is set by the EVG circuit's pulse width (190 ns), yielding a maximum achievable count rate of 5.3 MHz. Digital pulses corresponding to the channel's activity are shown in Fig. 3.1d.

3.3 Optoelectronic Characterization

We used a perimeter-gated SPAD fabricated in a 0.5-μm single-well, 2-metal, 3-poly CMOS process (shown schematically in Figs. 3.1a,b and 3.5) to investigate the effects of perimeter gating on DCR, SNR, and spectral responsivity. To investigate the effect of perimeter gating on PRNU, we used an imager including a 64 × 64 array of pg-SPAD pixels fabricated in a 0.35-μm CMOS process (Fig. 3.9).

3.3.1 Dark Count Rate

The DCR was estimated by counting the average number of pulses in 500 consecutively acquired frames of data, each lasting 10 ms (i.e., in a 100-Hz measurement bandwidth). The measurements were performed for gate voltages ranging from -10 to −16 V in steps of −0.5 V.

A total of 80 devices, extending over five chips, were tested. Figure 3.2a shows the DCR as a function of the perimeter gate voltage for 8 of the 80 channels characterized. The maximum DCR, measured at $V_G = -10$ V, was 218 kHz(±26 kHz), and the minimum achievable DCR was 1.3 kHz(±0.18 kHz) at $V_G = -16$ V. The area-normalized DCRs corresponding to the aforementioned figures were $\Gamma_{MAX} = 348\ \text{Hz} \cdot \mu\text{m}^{-2}$ and $\Gamma_{MIN} = 2.08\ \text{Hz} \cdot \mu\text{m}^{-2}$, respectively.

Figure 3.2b shows the yield of the sample lot. A device was considered to be nonfunctional if its DCR at $V_G = -16$ V exceeded 10^5 Hz. Using this criterion, the overall yield was 75%. Devices exhibiting a high DCR, in spite of the perimeter breakdown suppression, likely had a relatively larger defect density within the volumetric junction.

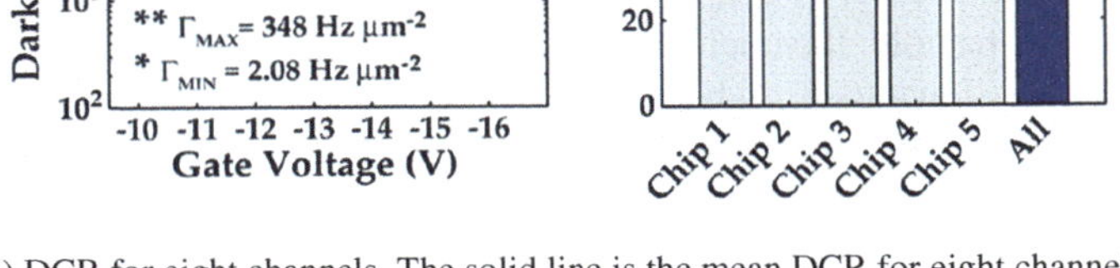

Fig. 3.2 (**a**) DCR for eight channels. The solid line is the mean DCR for eight channels from two different test chips. (**b**) Test-sample-lot yield. The asterisks indicate in which sample set the (∗) minimum and (∗∗) maximum area-normalized DCRs occurred

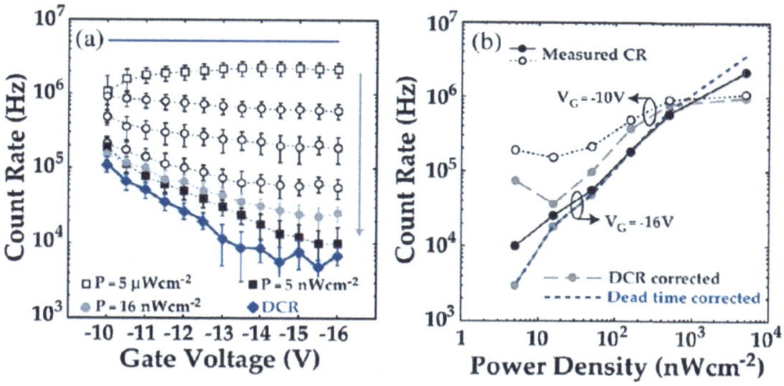

Fig. 3.3 (**a**) Count rate as a function of light intensity for a single channel. The arrow indicates the response for decreasing light intensity, and the horizontal line indicates the maximum achievable count rate. (**b**) Count rate as a function of optical power density for $V_G = -10$ V and for $V_G = -16$ V. Also shown are the DCR-corrected count rates (i.e., the mean DCR subtracted from the measured count rate under illumination) and the dead time (190 ns) corrected count rate for $V_G = -16$ V (see [19] for the canonical count rate correction procedure typically used in the literature)

3.3.2 Signal-to-Noise Ratio

A grating monochromator was used to provide a narrowband (2 nm) stimulus at 520 nm. The optical power of the stimulus was measured using a calibrated photometer, and the output beam of the monochromator was collimated onto a fiber bundle whose output was focused onto the test chip using a set of lenses.

In order to study the channels' response as a function of light intensity, neutral-density filters were used to attenuate the monochromator's output signal. The optical power density before attenuation was 5 μW/cm^2.

Figure 3.3a is a plot of a single channel's response under various illumination conditions, and Fig. 3.3b shows that each light level is readily discriminated when $V_G = -16$ V. This is not the case for $V_G = -10$ V. These results imply that larger magnitude gate voltages serve to increase the dynamic range by allowing lower light levels to be detected.

At $V_G = -16$ V and with 2.5 orders of magnitude attenuation of the stimulus, the signal remained distinguishable from the noise, i.e., its mean was larger than three standard deviations of the mean DCR. This puts the optical power density of the weakest signal detected in our experiments at 16 nW/cm^2 which corresponds to a photon arrival rate of 2.5×10^5 photons/s. The probability of detecting a photon was computed by dividing the measured count rate by the photon arrival rate; this calculation yielded a photon detection efficiency (PDE) of 10%.

The SNR was computed by dividing the DCR-corrected mean signal count rate by the mean DCR. Figure 3.4a shows the SNR as a function of V_G and for different illumination levels. Except when the signal was below the noise floor, the SNR always increased with increasing gate voltage magnitudes.

At high-illumination levels, the SNR showed an order of magnitude improvement (37-fold increase) over the gate voltage tuning range, whereas the improvement was

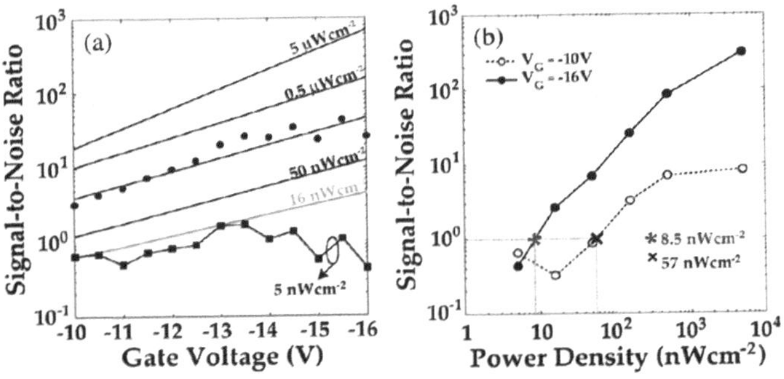

Fig. 3.4 (**a**) Exponential fits to the SNR data computed for five different light intensities. (For clarity, the fitted data are shown only for one trace.) (Black squares) Raw SNR data for an optical power density below the noise floor $\left(P = 5\text{nWcm}^{-2}\right)$ are also shown. (**b**) SNR as a function of optical power density for two gate voltages. At $SNR = 1$, ($\times$ and $*$) the optical power densities were extrapolated to find the minimum detectable signal for $V_G = -10$ V and $V_G = -16$ V

more moderate yet still appreciable at lower illumination levels (8-fold increase for a 16 nW/cm^2 stimulus).

Figure 3.4b shows the SNR for $V_G = -10$ V and for $V_G = -16$ V. The data show that lowering the perimeter gate voltage to -16 V reduces the extrapolated minimum detectable optical power density from 57 to 8.5 nW/cm^2, thereby maximizing the sensor's sensitivity.

Table 3.1 compares the pg-SPAD presented herein with published data from other CMOS SPAD architectures. The comparison is made on the basis of technology node, excess bias voltage V_E, measurement dead time t_D, DCR, area normalized DCR, PDE, noise-equivalent power (NEP), and active area. Here, the dead time refers to the time required for a complete quench and reset cycle; it may also include a hold-off period in the quench state (for suppressing afterpulsing) and the latency of the readout chain.

The devices reported make use of several PEB suppression techniques. Palubiak et al. [19], Marwick and Andreou [21], Niclass et al. [22], and Rochas et al. [23] all used diffused guard rings, and Finkelstein et al. [8], Gersbach et al. [20], and Richardson et al. [9] introduced techniques based on features available in modern process technologies, such as STI [8], passivated STI [20], and retrograde well doping [9]. As the table indicates, at $V_G = -16$ V, the pg-SPAD has an area-normalized DCR (Γ_{MIN}) that is comparable with that of state-of-the-art avalanche photodiodes [9, 21, 23] despite having the largest area.

3.3.3 *Spectral Responsivity*

In a SPAD, the photon detection efficiency (PDE) (a widely reported metric) is the ratio of the number of photoelectrons detected to the number of photons incident on

Table 3.1 Comparison with published submicrometer CMOS SPADs

Reference	CMOS Tech. (μm)	V_E (V)	t_D (ns)	DCR (Hz) ($T = 300°$K)	Γ_{MIN} (Hz$_\mu$ μm^{-2})	PDE (%)	NEP ($fW/\sqrt{Hz}$)	SPAD area (μm^2)
Palubiak et al. [19]	0.13	0.5	$\overline{12}$	1k	10	13(570 nm)	–	100
Richardson et al. [9]	0.13	0.8	–	47	0.9	18(440 nm)	–	50
Gersbach et al. [20]	0.13	1	180	90k	1.6k	–	–	58
Finkelstein et al. [8]	0.18	2.5	2	1M	26.3k	–	–	38
Marwick et al. [21]	0.18	0.5	–	100	1	14(670 nm)	–	79
Niclass et al. [22]	0.35	3.3	100	646	17	35(460 nm)	–	38
This work	0.50	1.15	190	218k($V_G = -10$ V)	348	14(520 nm)	35.6[a]	625
This work	0.50	1.15	190	$1.3k$($V_G = -16$ V)	2	10(520 nm)	5.30[a]	625
Rochas et al. [23]	0.80	2.5	< 10	60	2	21(440 nm)	0.025	30

[a] The NEP was estimated by scaling the unity SNR optical power densities by the active area and dividing by the square root of the measurement bandwidth (100 Hz)

the SPAD. The PDE depends on the probability of absorption of a photon and on the probability for the absorbed photon of triggering an avalanche. The probability of absorption depends in part on the spectral responsivity of the junction, and consequently, spectral responsivity is a key measurement when characterizing a SPAD.

$$R|_{\alpha V_B} = \begin{cases} \frac{|I_M - I_D|}{P_M}\left(\frac{A_P}{A_D}\right), \; for\ |I_M - I_D| > \delta \\ \epsilon, \; for |I_M - I_D| \leq \delta \end{cases} \tag{3.1}$$

3.3.3.1 Experimental Setup and Procedures

We defined the near-breakdown region as the voltage bias regime where the applied voltage across the junction ($V_{app} = V_C - V_A$ in Fig. 3.5) is nearly equal to the breakdown voltage V_B, i.e., equal to a voltage $\alpha \times V_B$, where $\alpha = 0.99$. The spectral responsivity R at $V_{app} = \alpha V_B$ is defined in Eq. 3.1, where P_M is the optical power of the incident light of wavelength λ, I_M is the photocurrent, I_D is the dark current of the device, A_P is the active area of the photometer head used to measure the optical power P_M, and A_D is the active area of the perimeter-gated SPAD. The

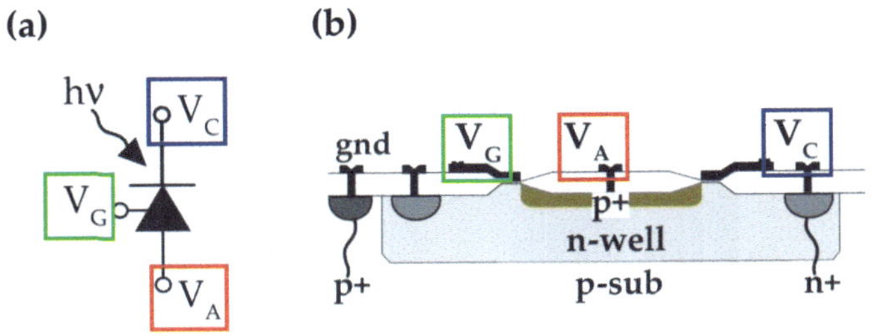

Fig. 3.5 (**a**) Circuit schematic of a pg-SPAD under test. (**b**) Cross-sectional view of a perimeter-gated single photon avalanche diode (pg-SPAD). (Adapted from [2])

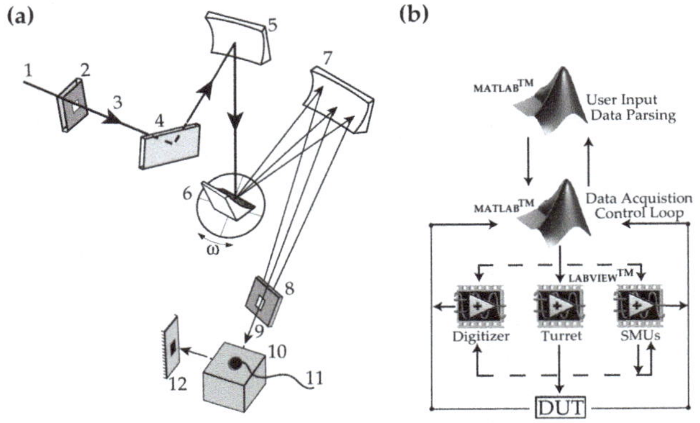

Fig. 3.6 Experimental setup for spectral characterization. (**a**) Monochromator: (1) Hg-Xe light source; (2) entrance slit; (3) white light beam; (4), (5), and (7), concave mirrors; (6) motorized turret and holographic grating assembly; (8) exit slit; (9) monochromatic light beam; (10) integrating sphere; (11) calibrated photometer head; and (12) device under test (DUT). The digitizer, the monochromator turret, and the source-measure units (SMUs) are controlled by Labview virtual instruments. Higher-level functions such as user input, data parsing, and data acquisition control are achieved using Matlab routines. (**b**) Software architecture utilizing custom-written Matlab TM and Labview^{TM}

parameter δ is chosen to be a small number (e.g., 1 standard deviation of the average dark current at V_{app}), and the parameter ϵ is nearly zero, indicating that there is no distinction between the measured dark current and photocurrents. The parameter ϵ is chosen to be the floating point accuracy number *eps* in Matlab in order to allow plotting the responsivity on a log scale.

The instrumentation for measuring the spectral responsivity is shown in Fig. 3.6a. The system includes a grating monochromator equipped with a broadband light source, a calibrated optical power meter, and data acquisition hardware. The monochromator's output signal is coupled to an integrating sphere whose interior is coated with a diffuse reflector. This means that the optical signals at the sphere's output ports are spatially uniform and equal in magnitude. This facilitates alignment and ensures reproducible data. Further, the monochromator's output port includes a

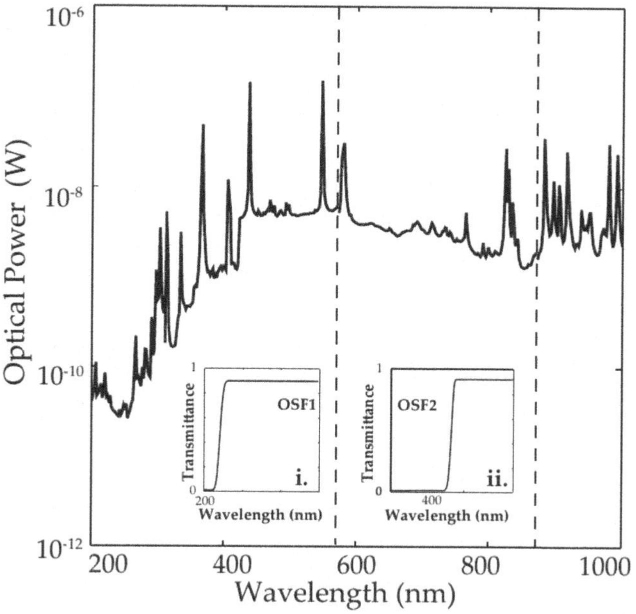

Fig. 3.7 Hg-Xe lamp spectrum, obtained with order sorting filters (OSFs) 1 and 2 (inset i and ii). (The Matlab TM logo is a registered trademark of Mathworks, Inc., Boston, Massachusetts, USA. The Labview TM VI icon is a registered trademark of National Instruments, Inc., Austin, Texas, USA)

shutter system allowing one to measure the dark current I_D when the shutter is closed and the photocurrent I_M when the shutter is opened.

The device under test (DUT), i.e., the perimeter-gated SPAD, and a calibrated photometer head are placed orthogonally at the sphere's output ports. The system is computer-controlled, with data acquisition control implemented using Matlab and lower-level instrument control, including instruments for actuating the grating's turret, implemented using Labview virtual instruments (Fig. 3.6b). A more detailed description of the system is provided in the caption of Fig. 3.6. Figure 3.7 shows the optical signal at the sphere's output ports, measured with the calibrated photometer head.

Using the system of Fig. 3.6, we estimated the spectral responsivity of a 35-μm square p+n-well junction having chamfered corners and a poly-1 perimeter gate [1]. The responsivity was measured using Algorithm 1, which lists the main operations and measurements carried out by the system shown in Fig. 3.6. Algorithm 1 begins by first setting several sweep variables and parameters. These include the applied voltage to be sourced across the pg-SPAD (line 2), the gate voltage to be sourced to the perimeter gate (line 3), and the wavelength range to be outputted from the monochromator into the integrating sphere (line 4).

Algorithm 1 Responsivity measurement algorithm

```
procedure ESTIMATE R|V_G(I_M, I_D, P_M(λ))
V_app ← V_A1 : step: V_Am
V_G ← V_G1 : step: V_Gn
λ ← λ_1 : step : λ_j
for i = 1 : n do
    close shutter
    source V_G(i)
    sweep V_app, measure I_D
    compute V_B {See ref. [10]}
    V_A(m−k) ≈ α V_B, α = 0.99 and k < m
    V*_A ← V_A1: step: V_A(m−k)
    I*_D ← I_D(V*_A)
    open shutter
    for r = 1 to j do
        source V_G(i)
        sweep V*_A, measure I_M
        compute R|V_G(i)(I_M, I*_D, P_M(λ_r))
    end for
end for
return R
close shutter
end procedure
```

Algorithm 1 then enters a measurement loop that includes two nested for loops. The first loop measures the dark current of the pg-SPAD at each gate voltage V_G. The second loop measures the photocurrent at each set wavelength λ and for each gate voltage V_G and across the range of applied voltages. Various near-breakdown spectral responsivities were obtained and are shown in Fig. 3.8.

3.3.3.2 Results

The spectral responsivity of the pg-SPAD was estimated for gate voltage magnitudes ranging from 0 to 16 V in steps of 2 V. The reverse bias voltage, i.e., the applied voltage V_{app}, was swept from about 10 V to the near-breakdown voltage, i.e., to the reverse bias voltage that satisfies $V_{app} = \alpha \times V_B$.

As previously reported [1, 10, 18], increasing the gate voltage's magnitude increases the breakdown voltage of the pg-SPAD. Therefore, for each gate voltage sourced, the breakdown voltage was different and, thus, so was the near-breakdown voltage condition. As such, for each gate voltage sourced, a different applied voltage range was obtained, as indicated by the different ranges of reverse bias voltages shown on the y-axes in the color plots of Fig. 3.8.

The results indicate that spectral responsivity is strongly influenced by the gate voltage. However, while it would appear that the pg-SPAD is highly responsive to short wavelengths, this is likely not the case. Interestingly, for example, for low-magnitude gate voltages (Fig. 3.8a), the pg-SPAD exhibits what appears to be

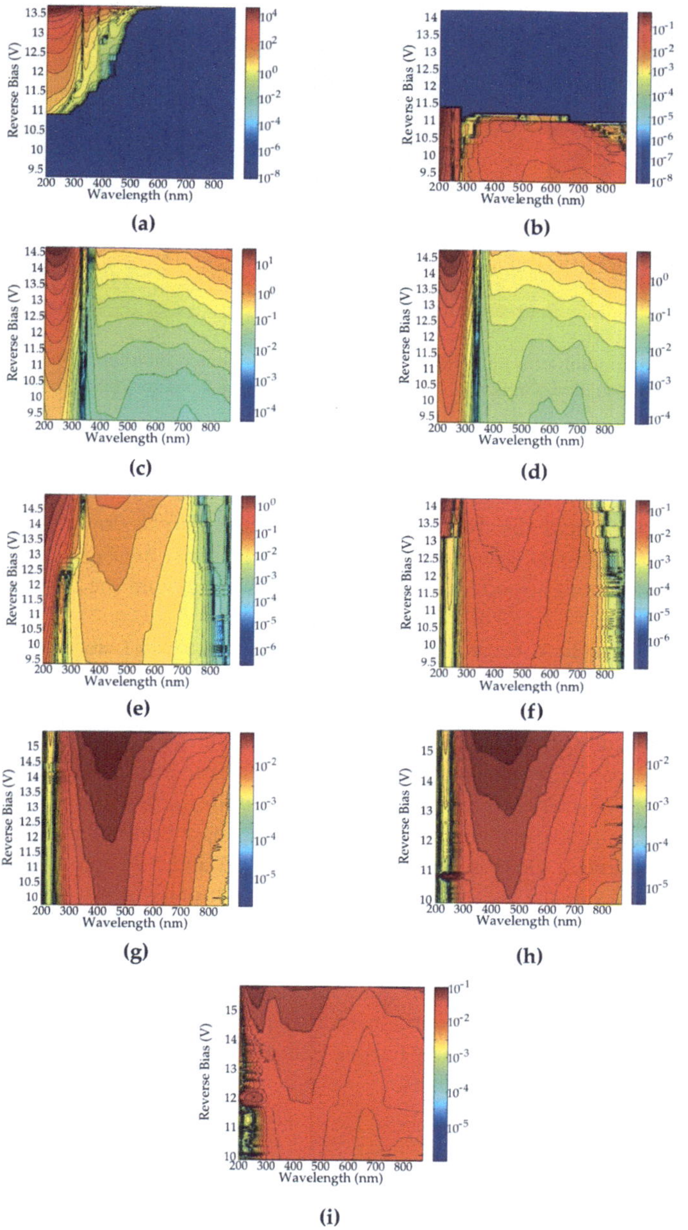

Fig. 3.8 Near-breakdown spectral responsivity using Algorithm 1. The color bar units are amperes per watts (A/W). (**a**) $V_G = 0$ V. (**b**) $V_G = -2$ V. (**c**) $V_G = -4$ V. (**d**) $V_G = -6$ V. (**e**) $V_G = -8$ V. (**f**) $V_G = -10$ V. (**g**) $V_G = -12$ V. (**h**) $V_G = -14$ V. (**i**) $V_G = -16$ V

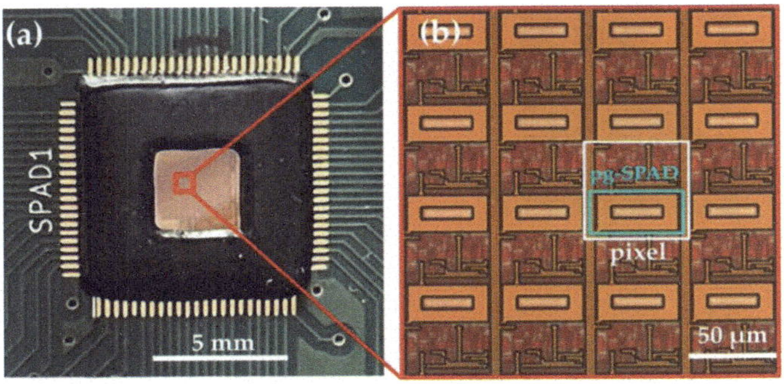

Fig. 3.9 (**a**) Photograph of a pg-SPAD imager prototype mounted on a test printed circuit board (PCB), with encapsulated wirebonds. (**b**) Photomicrograph of a 4 × 4 section of the pg-SPAD pixel array

a large response between 200 nm and 400 nm. However, as we have previously demonstrated, at these low biasing regimes, the edges break down first as a result of perimeter edge breakdown [10]. As such, much of the measured current is due to edge effects, in addition to the photocurrent resulting from the absorption of the incident photons.

The results further reveal that as the gate voltage increases in magnitude, the spectral responsivity changes markedly. For instance, at $V_G = 0$, the pg-SPAD was largely unresponsive to longer wavelengths. However, once the gate voltage magnitude increased to 2 V and above, the pg-SPAD showed a much broader responsivity profile over the visible range. Furthermore, as the gate voltage increased, the spectral responsivity became more uniform over the visible range, as shown in Fig. 3.8c,d,e, and f. Above a gate voltage magnitude of 10 V, the spectral responsivity showed clear maxima in between 400 and 590 nm, consistent with typical peak regions for the responsivity of p+/n-well junctions.

Another interesting result is that from $V_G = -12$ V to $V_G = -14$ V (Fig. 3.8g and Fig. 3.8h), the responsivity spectrum was similar in magnitude and profile, despite the application of different gate voltages. This suggests that there exists a gate voltage range where the spectral responsivity remains substantially unchanged. Such a feature provides the ability to tune gate voltages to alter breakdown without compromising spectral responsivity.

3.3.4 Photoresponse Nonuniformity

In this section, we present optoelectronic characterization results from an imager including a 64 × 64 array of pg-SPAD pixels (Fig. 3.9). Each pixel includes avalanche sensing circuitry and connects to peripheral hardware configured for an all-digital readout scheme.

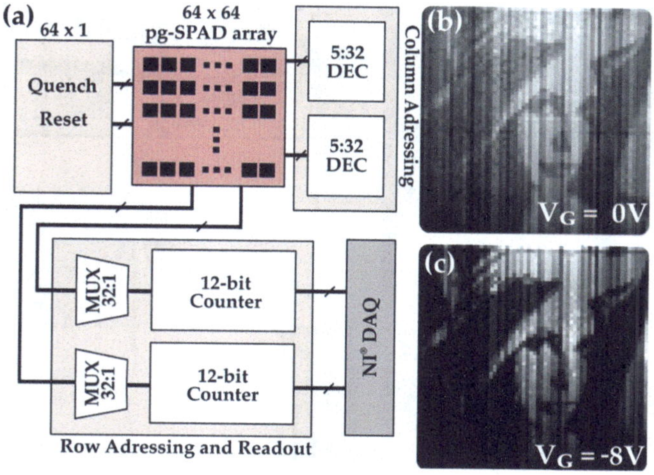

Fig. 3.10 (**a**) Block diagram of the pg-SPAD imager. The peripheral circuitry includes decoders for column addressing and multiplexed counters for registering pixel counts. Quenching and reset are provided row-wise using an active quenching and reset circuit. For circuit architecture of the pg-SPAD imager refer to [24]. (**b**) and (**c**) Photographs taken by the pg-SPAD imager at two different perimeter gate voltages

The imager's block diagram is shown in Fig. 3.10a. The architecture includes row-wise active-quenching and reset circuits, which are interfaced with a selected pixel that outputs a digital pulse whose leading edge coincides in time with an avalanche event occurring in the pg-SPAD. Two pixels are read out simultaneously using a National Instruments ® data acquisition (DAQ) system connected to a PC. Signal values are discussed in [24].

Figure 3.10b and c shows two photographs taken by the imager, each for a different perimeter gate voltage V_G. To take these photographs, the imager chip was fitted with a lens, and the scene consisted of a black and white image printed on a transparency sheet that was taped to a back-illuminated screen. As observed from the photographs, there is a fixed pattern noise (FPN) [25] component that is readily apparent. The FPN component includes a dark signal component, a readout component, and a component that is due to differences in photoresponse at each pixel; this latter component is the imager's photoresponse nonuniformity (PRNU) [26, 27].

Because the PRNU depends on pixel spectral response and perimeter gating is intricately related to changes in spectral responsivity [28, 29], we studied the imager's spectral PRNU as a function of perimeter gating. Our experimental results show that the single-pixel spectral responsivity occupies two regimes, one in which small perimeter gate voltage magnitudes appear to cause short wavelength responsivity to dominate and another in which large perimeter gate voltage magnitudes cause the device to have a typical responsivity profile like the ones shown in ref. [30].

Fig. 3.11 Optical setup for measuring the pg-SPAD imager's optoelectornics characteristics. (1) Laser-driven light source (LDLS). (2) Lamp housing and coupler. (3) Triple-grating monochromator. (4) Filter wheel preloaded with neutral-density filters. (5) Collimating lens. (6) The pg-SPAD imager under test. (7) Data acquisition (DAQ) board. (8) Optical power meter with a silicon photodiode coupled to a 40-μm pinhole. (9) Power supplies

Our experimental results further show that the average PRNU and its standard deviation are large in the short wavelength-dominated responsivity regime, whereas both measures are significantly reduced in the latter regime. These results suggest that large perimeter gating voltages facilitate PRNU extraction in addition to being more favorable for typical PRNU removal methods used in image correction [3].

3.3.4.1 Experimental Setup and Procedures

Figure 3.11 shows our experimental setup. Monochromatic light was provided by a triple-grating monochromator (Horiba, iHR320) coupled to a broadband laser-driven light source (Energetiq, LDLS EQ-99X) using a custom-designed coupler. The monochromator's internal 550-nm long pass filter was placed in the optical path when the selected grating was positioned to output light in the 580 nm–1100 nm range; this was done to remove higher-order harmonics generated by the grating. Furthermore, the monochromator's entrance and exit slit widths were set to 7 mm and 2 mm, respectively, to provide a nominal spectral resolution of 2 nm.

Further, a twelve-position computer-controlled motorized filter wheel preloaded with neutral-density filters (Thorlabs, FW212CNEB) was placed at the monochromator's exit port. This was done to minimize wide swings in optical power inherent to the output light's power spectrum (Fig. 3.12). After correction with the filter

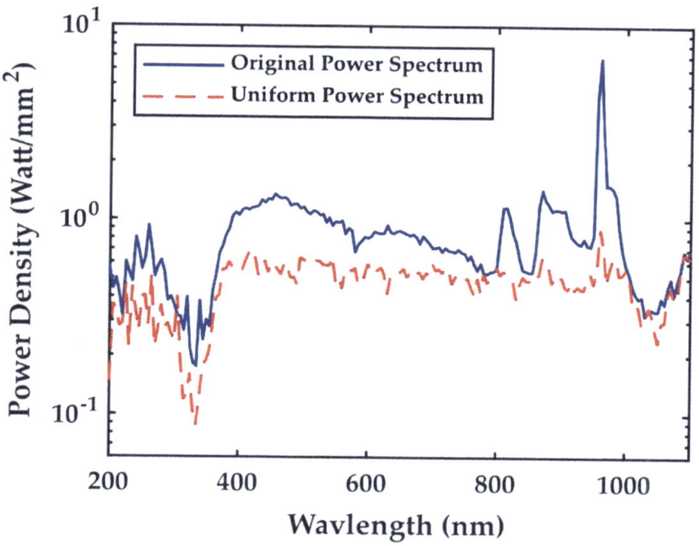

Fig. 3.12 Optical power density spectrum without correction (solid line) and filter wheel-corrected optical power density spectrum (dashed line). Power spectrum uniformity is achieved in the visible range

wheel, the output light was passed through a collimating lens before reaching the imager. To maintain parallel incident rays and homogeneous illumination on the chip, the imager was placed at a distance equal to one focal length from the filter wheel.

A calibrated reference detector fitted with a 40-μm pinhole (Thorlabs, S120VC, P40K) and an optical power meter (Thorlabs, PM100D) were used to measure the power spectra shown in Fig. 3.12. During measurements, the whole setup was enclosed in a dark box to prevent contamination from ambient light. The monochromator, the optical power meter, the neutral-density filter wheel, and the DAQ system were controlled via a single Visual Studio ®C + +/ CLI project.

3.3.4.2 Single-Pixel Spectral Responsivity Measurement

The integration time for each pixel was chosen to be 100 ms, and the spectral responsivity of a pixel located at the center of the array was measured as a function of perimeter gate voltage, i.e., scanning V_G from 0 V to -8 V in steps of -1 V. The V_G values were picked based on our previous results [2, 24]. The incident light wavelength was scanned from $\lambda = 200$ nm to $\lambda = 1100$ nm in 5 nm steps. Here, the spectral responsivity was expressed as the ratio of the number of registered counts per second to the optical power incident on the pg-SPAD's active area. Algorithm 2 was used with our setup to measure the responsivity for the different perimeter gate voltages noted above. The measured responsivity spectra are shown in Fig. 3.13.

Algorithm 2 Single-pixel responsivity measurement algorithm

procedure ESTIMATE $R|_{VG}(P_M(\lambda))$
$\lambda \leftarrow \lambda_1$: step : λ_j
$V_G \leftarrow V_{G1}$: step : V_{Gn}
close slits
for i = 1 : n **do**
    source $V_G(i)$
    measure dark count
end for
open slits
for r = 1 : n **do**
    source $V_G(r)$
    for k = 1 : j **do**
        set filter wheel to its proper position
        set wavelength to λ_j
        measure photon count
        if photon count $\geq$ dark count **then**
            photon count = photon count - dark count
        else
            photon count = 0
        end if
        return photon count
    end for
end for
close slits
calculate $R|_{VG}(P_M(\lambda)) = \dfrac{PC/t_i}{P_M(\lambda)}$
end procedure

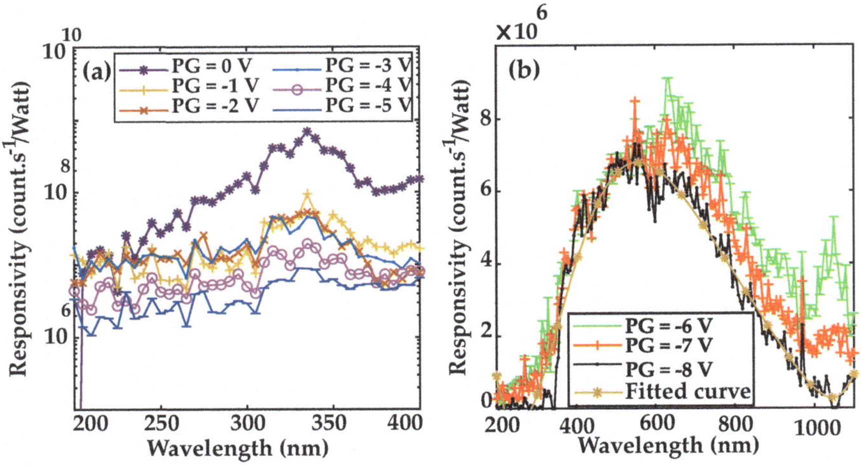

Fig. 3.13 (**a**) Measured spectra where wavelengths in the 200–400 nm range have the largest responsivity. This is achieved using $V_G = 0$ V to -5 V. (**b**) Measured spectra in the 200–1100 nm range for $V_G = -6$ V, -7 V, and -8 V. A 6th degree polynomial fit showed maximum responsivity of 6.8 $\times 10^6$ counts. s^{-1}/W at λ = 545 nm

Algorithm 3 PRNU calculation algorithm

procedure ESTIMATE $PRNU|_{V_G}(\lambda)$
$V_G \leftarrow V_{G1}$: step: V_{Gn}
$\lambda \leftarrow \lambda_1$: step : λ_j
close slits
for k = 1 : n **do**
  source V_G(k)
  obtain high-frequency dark frame $n_{hf_d}(k)$
  $DSNU(k) = \sigma(n_{hf_d}(k))$
end for
for r = 1 : j **do**
  for k = 1 : n **do**
    source $V_G(k)$
    obtain high-frequency illuminated frame $n_{hf_i}(r, k)$
    calculate $PRNU(r, k)$
  end for
end for
end procedure

3.3.4.3 Photoresponse Nonuniformity Measurement

As previously noted, the imager's FPN includes multiple components. In order to estimate the imager's PRNU, the component that is due to the pixels' DCR, i.e., the dark signal nonuniformity (DSNU), must first be measured. Furthermore, the readout component must also be removed from the acquired images.

We adapted a method proposed by Kim et al. to estimate the DSNU, the PRNU, and filter out low-frequency noise components corresponding to readout fixed pattern noise [31]. To estimate the DSNU, we first closed the monochromator's shutter to keep the imager in the dark. The pixel array was scanned five times, yielding five dark frames $f_1, f_2, \ldots, f_5$. An average frame $\bar{f}$ was obtained from the five frames in order to reduce any temporal noise.

The method consisted of first applying a Wiener filter [32] on $\bar{f}$, yielding $DNF(\bar{f})$. A residual noise frame n was then obtained by subtracting $DNF(\bar{f})$ from $\bar{f}$. The DSNU is a high-frequency signal contained in n, whereas the low-frequency components are centered at low frequencies in n.

Thus, a high-frequency noise frame n_{hf_d} was obtained by zeroing the low-frequency coefficients in the discrete cosine transform (DCT) of n [33]. Specifically, n_{hf_d} was obtained using $DCT^{-1}(DCT(n \circ m))$, where $\circ$ represents the entry-wise product operator and m is a high-pass filter mask. Illuminated high-frequency noise frames n_{hf_i} were similarly obtained when the imager was subjected to light from the monochromator. Example results from the abovenoted procedures are shown in Fig. 3.14.

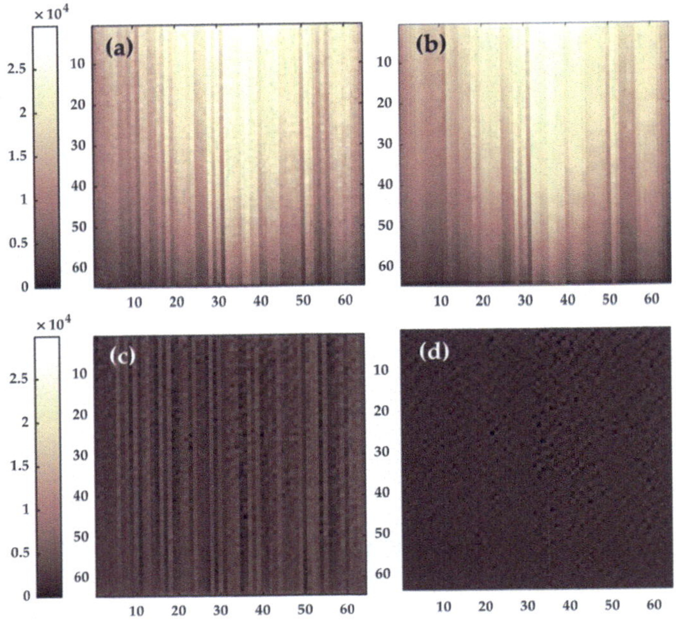

Fig. 3.14 Example of the steps undertaken when obtaining the high-frequency noise frame for $V_G = 0$ V and $\lambda = 400$ nm. (**a**) Acquired illuminated frame. (**b**) Low-frequency noise frame extracted by Wiener filtering. (**c**) Residual noise frame n. (**d**) High-frequency noise frame n_{hf_i}

$$\sigma(n_{hf_d}) = \sqrt{\sum_{i=1}^{64}\sum_{j=1}^{64}(n_{hf_{dij}} - \mu(n_{hf_d}))^2} \tag{3.2}$$

$$PRNU = \frac{\sqrt{|\sigma^2(n_{hf_i}) - \sigma^2(n_{hf_d})|}}{|\mu(n_{hf_i}) - \mu(n_{hf_d})|} \times 100\% \tag{3.3}$$

A single metric characterizing the spatial DSNU was defined as the standard deviation (σ) of the distribution of values across the pixel array for the high-frequency dark noise frame, n_{hf_d} (Eq. 3.2). Similarly, a single metric was obtained for the illuminated high-frequency noise frames $\left(n_{hf_i}\right)$. Both metrics were calculated for frames obtained across the wavelength and perimeter gate voltage scan vectors. Lastly, the PRNU was calculated as shown in Eq. 3.3, similarly to the calculation prescribed in the EMVA camera and image sensor characterization standard [34] and using Algorithm 3. The PRNU measurement results are shown in Fig. 3.15.

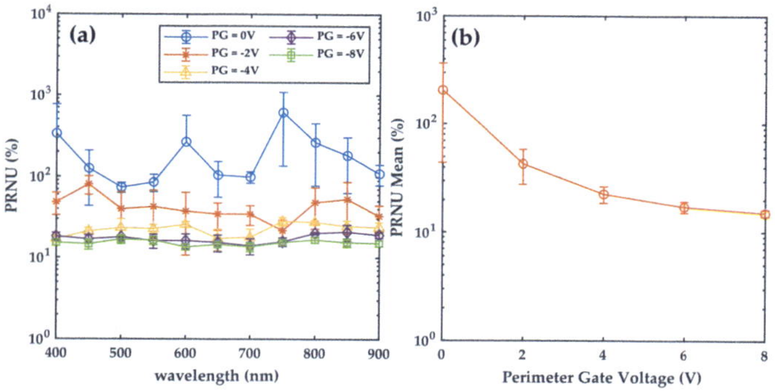

Fig. 3.15 (**a**) The imager's estimated PRNU for λ = 400 nm–900 nm and for V_G = 0 V to −8 V. Three full array measurements were performed for each data point. (**b**) Mean of the pg-SPAD imager's estimated PRNU across all wavelengths as a function gate voltage

3.3.4.4 Results

The spectral responsivity measurements of Fig. 3.13 revealed that there were two regimes of photoresponse. The first regime ($V_G = 0$ V to $V_G = 5$ V in magnitude) consisted of high responsivities experienced in the short wavelength region (200 nm–400 nm); these responsivities were elicited by relatively smaller magnitude perimeter gate voltage. The second regime (VG = 6 V to VG = 8 V in magnitude) yielded bell-shaped responsivity curves for the whole test range (200 nm–1100nm).

We caution that the results obtained for the first regime do not entail that perimeter gating makes the device perform well as an ultraviolet imager. Rather, this apparent increase in photoresponse is likely the result of excessive dark count generation rates that are proportional to the number of photogenerated carriers (e.g., afterpulsing). And, because these dark count events are initiated as a result of light absorption, particularly in regions where PEB has not been completely curtailed, merely subtracting a previously acquired dark frame from the image is not sufficient to eliminate the dark noise contribution.

Furthermore, as shown in Fig. 3.15a for the first regime, i.e., when relatively small perimeter gate voltages are applied, the uncertainty in the PRNU is quite large, which suggests that the PRNU is heavily influenced by the device's noise. In contrast, the uncertainty in the PRNU is much reduced when perimeter gate voltages are large in magnitude, i.e., when the device is in the second responsivity regime. Figure 3.15b shows that with increasing perimeter gate voltage magnitude, the average estimated PRNU across all wavelengths decreased from 207.05% with $\sigma = 162.50\%$ to 15.32% with $\sigma = 1.11\%$.

3.4 Conclusions

In this chapter, we showed that perimeter field gating reduced a SPAD's DCR. For high illumination, it maximized SNR, and for low illumination, it maximized sensitivity. Since the DCR increases strongly with excess bias voltage (V_E) and depends on the process technology, performance in specific applications will depend on the PDE and dead time (t_D). On that basis, the device reported by Rochas et al. [23] had the best performance due to its low DCR at relatively large V_E and short dead time. However, it was fabricated in an older CMOS process, limiting array density and fill factor. As fabrication technologies scale down, the trade-offs between DCR and excess bias voltage become significant; this explains the relatively higher DCR observed at smaller V_E for the other devices listed in Table 3.1.

Although STI techniques preserve fill factor, STI-based devices [8, 20] exhibit the highest DCR, suggesting that fill factor is improved at the expense of noise. Conversely, our data showed a $\sim$ 7-fold reduction in NEP using perimeter gating, a technique which also preserves fill factor. This suggests that a field-based PEB suppression method is viable for counteracting the degradation in noise performance observed in deep submicrometer CMOS technologies.

We also reported the effects of perimeter gating on the spectral responsivity of a SPAD. Experiments showed that perimeter gating strongly influenced spectral responsivity by broadening the response to visible light. This suggests that perimeter gating effectively suppresses PEB and allows the volumetric junction of the measured devices to enter breakdown and contribute to device current.

Lastly, we investigated the spectral responsivity and the PRNU of a 64×64 array of pg-SPAD imager as a function of perimeter gating. Our results revealed that perimeter gating serves to reduce the uncertainty in the device's PRNU across all wavelengths. We found that noise generation significantly contributed to the PRNU as well as to its uncertainty for small-magnitude perimeter gate voltages. In contrast, for large-magnitude perimeter gate voltages, the PRNU and the uncertainty in its measurement were significantly reduced, suggesting a diminished dependence on device noise. These results imply that perimeter gating can further serve to reduce photoresponse nonuniformity of Geiger-mode imagers.

References

1. M. Dandin, P. Abshire, High signal-to-noise ratio avalanche photodiodes with perimeter field gate and active readout. IEEE Electron Dev. Lett. **33**(4), 570–572 (2012). https://doi.org/10.1109/LED.2012.2186112
2. M. Dandin, P. Abshire, Near Breakdown Spectral Responsivity of Perimeter-Gated Single-Photon Avalanche Diodes, in *Midwest Symposium on Circuits and Systems*, vol. 2017, pp. 867–870 (August 2017). https://doi.org/10.1109/MWSCAS.2017.8053061. ISBN:9781509063895. ISSN:15483746

3. F. Dehghandehnavi, M.S. Sajal, K.C. Lin, M. Dandin, Spectral Responsivity and Photoresponse Non-uniformity of a Perimeter-Gated Single-Photon Avalanche Diode Imager, in *Midwest Symposium on Circuits and Systems* (Institute of Electrical and Electronics Engineers Inc., 2023), pp. 317–321. https://doi.org/10.1109/MWSCAS57524.2023.10405979. ISBN:9798350302103. ISSN:15483746
4. S. Sze, G. Gibbons, Effect of junction curvature on breakdown voltage in semiconductors. Solid State Electron. **9**(9), 831–845 (1966). https://doi.org/10.1016/0038-1101(66)90033-5
5. V. Temple, M. Adler, Calculation of the diffusion curvature related avalanche breakdown in high-voltage planar p-n junctions. IEEE Trans. Electron Dev. **22**(10), 910–916 (1975). https://doi.org/10.1109/T-ED.1975.18241
6. A. Pauchard, P.-A. Besse, R.S. Popovic, Simulation of a New CMOS Compatible Method to Enhance the Breakdown Voltage of Highly Doped Shallow PN Junctions, in *Modeling and Simulation of Microsystems*, pp. 420–425 (1998)
7. A. Rochas, A.R. Pauchard, P.A. Besse, D. Pantic, Z. Prijic, R.S. Popovic, Low-noise silicon avalanche photodiodes fabricated in conventional CMOS technologies. IEEE Trans. Electron Dev. **49**(3), 387–394 (2002). https://doi.org/10.1109/16.987107
8. H. Finkelstein, M.J. Hsu, S.C. Esener, STI-bounded single-photon avalanche diode in a deep-submicrometer CMOS technology. IEEE Electron Dev. Lett. **27**(11), 887–889 (2006). https://doi.org/10.1109/LED.2006.883560
9. J.A. Richardson, E.A. Webster, L.A. Grant, R.K. Henderson, Scaleable single-photon avalanche diode structures in nanometer CMOS technology. IEEE Trans. Electron Dev. **58**(7), 2028–2035 (2011). https://doi.org/10.1109/TED.2011.2141138
10. M. Dandin, A. Akturk, B. Nouri, N. Goldsman, P. Abshire, Characterization of single-photon avalanche diodes in a 0.5 μm standard CMOS process - Part 1: Perimeter breakdown suppression. IEEE Sens. J. **10**(11), 1682–1690 (2010). https://doi.org/10.1109/JSEN.2010.2046163
11. F. Zappa, S. Tisa, A. Tosi, S. Cova, Principles and features of single-photon avalanche diode arrays. Sens. Actuators A Phys. **140**(1), 103–112 (2007). https://doi.org/10.1016/j.sna.2007.06.021
12. E. Kamrani, F. Lesage, M. Sawan, Premature Edge Breakdown Prevention Techniques in CMOS APD Fabrication, in *IEEE 10th International New Circuits and Systems Conference*, pp. 345–348 (2012). https://doi.org/10.1109/NEWCAS.2012.6329027. ISBN:9781467308595
13. M. Sicre, M. Agnew, C. Buj, J. Coignus, D. Golanski, R. Helleboid, B. Mamdy, I. Nicholson, S. Pellegrini, D. Rideau, D. Roy, F. Calmon, Dark Count Rate in Single-Photon Avalanche Diodes: Characterization and Modeling Study, in *ESSDERC 2021 - IEEE 51st European Solid-State Device Research Conference (ESSDERC)* (IEEE, 2021), pp. 143–146. https://doi.org/10.1109/ESSDERC53440.2021.9631797
14. R.B. Merrill, Color separation in an active pixel cell, imaging array using a triple well structure, *U.S. Patent 5,965,875*, 10 1999
15. D. Ho, M.O. Noor, U.J. Krull, G. Gulak, R. Genov, CMOS tunable-wavelength multi-color photogate sensor. IEEE Trans. Biomed. Circuits Syst. **7**(6), 805–819 (2013). https://doi.org/10.1109/TBCAS.2013.2243727
16. F.D.S. Campos, N. Faramarzpour, O. Marinov, M.J. Deen, J.W. Swart, Photodetection with gate-controlled lateral BJTs from standard CMOS technology. IEEE Sens. J. **13**(5), 1554–1563 (2013). https://doi.org/10.1109/JSEN.2012.2235827
17. Y. Maruyama, K. Sawada, H. Takao, M. Ishida, The fabrication of filter-less fluorescence detection sensor array using CMOS image sensor technique. Sens. Actuators A Phys. **128**(1), 66–70 (2006). https://doi.org/10.1016/j.sna.2006.01.017
18. M. Dandin, M.H.U. Habib, B. Nouri, P. Abshire, N. McFarlane, Characterization of single-photon avalanche diodes in a 0.5-μm standard CMOS process-part 2: equivalent circuit model and geiger mode readout. IEEE Sens. J. **16**(9), 3075–3083 (2016). https://doi.org/10.1109/JSEN.2016.2526665
19. D. Palubiak, M.M. El-Desouki, O. Marinov, M.J. Deen, Q. Fang, High-speed, single-photon avalanche-photodiode imager for biomedical applications. IEEE Sens. J. **11**(10), 2401–2412 (2011). https://doi.org/10.1109/JSEN.2011.2123090

20. M. Gersbach, C. Niclass, E. Charbon, J. Richardson, R. Henderson, L. Grant, A Single Photon Detector Implemented in a 130 nm CMOS Imaging Process, in *38th European Solid-State Device Research Conference* (IEEE, Edinburgh, UK, 2008), pp. 270–273. https://doi.org/10.1109/ESSDERC.2008.4681750. ISBN:9781424423644
21. M.A. Marwick, A.G. Andreou, Single photon avalanche photodetector with integrated quenching fabricated in TSMC 0.18 μm 1.8 v CMOS process. Electron. Lett. **44**(10), 643–644 (2008). https://doi.org/10.1049/el:20080673
22. C. Niclass, C. Favi, T. Kluter, M. Gersbach, E. Charbon, A 128 ×128 single-photon imager with on-chip column-level 10 b time-to-digital converter array capable of 97 ps resolution. IEEE J. Solid State Circuits **43**(12), 2977–2989 (2008). https://doi.org/10.1109/JSSC.2008.20064
23. A. Rochas, P.A. Besse, R.S. Popovic, Actively recharged single photon counting avalanche photodiode integrated in an industrial CMOS process. Sens. Actuators A Phys. **110**(1–3), 124–129 (2004). https://doi.org/10.1016/j.sna.2003.08.003
24. M.S. Sajal, K.C. Lin, B. Senevirathna, S. Lu, M. Dandin, Perimeter-Gated Single-Photon Avalanche Diode Imager with Vanishing Room Temperature Dark Count Probability, in *ICECS 2022–29th IEEE International Conference on Electronics, Circuits and Systems, Proceedings* (Institute of Electrical and Electronics Engineers Inc., 2022). https://doi.org/10.1109/ICECS202256217.2022.9970824. ISBN 9781665488235
25. E.D.S. Pereira, Determining the fixed pattern noise of a CMOS sensor: improving the autonomous star trackers sensibility. J. Aerospace Technol. Manag. **5**(2), 217–222 (2013). https://doi.org/10.5028/jatm.v5i2.206
26. T. Zhang, X. Li, J. Li, Z. Xu, CMOS fixed pattern noise removal based on low rank sparse variational method. Appl. Sci. **10**(11), 3694 (2020). https://doi.org/10.3390/app10113694
27. M. Klosowski, Y. Sun, Fixed pattern noise reduction and linearity improvement in time-mode CMOS image sensors. Sensors **20**(20), 5921 (2020). https://doi.org/10.3390/s20205921
28. C. Ye, D.-h. Lee, CMOS Image Sensor: Characterizing Its PRNU (Photo-Response Non-uniformity), in *Optical Data Storage 2018: Industrial Optical Devices and Systems*, p. 10, ed. by R. Katayama, Y. Takashima (2018). https://doi.org/10.1117/12.2321168. ISBN:9781510620858
29. S. Knight, S. Moschou, M. Sorell, Analysis of Sensor Photo Response Non-uniformity in RAW Images, in *Forensics in Telecommunications, Information and Multimedia*, pp. 130–141 (2009). https://doi.org/10.1007/978-3-642-02312-5_15
30. R. Kuroda, S. Sugawa, Cameras with On-Chip Memory CMOS Image Sensors, in *The Micro-World Observed by Ultra High-Speed Cameras* (Springer International Publishing, Cham, 2018), pp. 103–124. https://doi.org/10.1007/978-3-319-61491-5_5
31. Y. Kim, Y. Lee, CamPUF: Physically Unclonable Function Based on CMOS Image Sensor Fixed Pattern Noise, in *55th Annual Design Automation Conference* (2018). https://doi.org/10.1145/3195970.3196005
32. J.S. Lim, *Two-Dimensional Signal and Image Processing* (Prentice Hall, Englewood Cliffs, NJ, 1990), p. 548, equations 9.44, 9.45, and 9.46
33. A.K. Jain, *Fundamentals of Digital Image Processing*, 1st edn. (Prentice Hall, Englewood Cliffs, NJ, 1988)
34. European Machine Vision Association, EMVA Standard 1288: Standard for Characterization of Image Sensors and Cameras (2010). www.emva.org

Chapter 4
Perimeter-Gated Single-Photon Avalanche Diode Imagers

4.1 Introduction

Single-photon avalanche diode (SPAD) imagers have been realized in complementary metal oxide semiconductor (CMOS) processes as two-dimensional arrays interfaced with monolithically integrated readout and processing circuits. These imagers are typically used in synchrony with an active light source such as a pulsed laser. In this mode of operation, a count registered by the device can be temporally correlated with known photon arrival times. This provides a means to distinguish photon counts from spurious counts that originate from several noise sources inherent to the device and to its Geiger-mode operation. This active imaging approach is specific to applications where photon arrival times are known *a priori* [1–4]. Example applications that utilize this measurement paradigm include fluorescence lifetime imaging microscopy (FLIM) [5, 6] and light detection and ranging (LiDAR) [7, 8].

Several groups have recently investigated the application of SPAD imagers in the detection of high photon fluxes using a passive imaging mode, i.e., with no temporal synchronization to a photon source [9–11]. In this scenario, SPAD imagers were shown to have several advantages over their conventional counterparts.

On the one hand, SPAD readout circuits can be made much faster compared to that of charge-coupled devices (CCDs), electron multiplication (EM) CCDs, or active pixel sensor (APS) imagers. Since an avalanche current pulse provides a measurable signal without needing any intervening amplification, a SPAD pixel's single-photon sensitivity and its high temporal resolution enable much higher frame rates than what is currently possible with typical imagers [12, 13].

On the other hand, a SPAD's photon counting capability translates to an increased dynamic range relative to the dynamic range of a conventional APS-based imager pixel. Specifically, APS pixels exhibit a hard saturation when they reach full-well capacity. In contrast, SPAD pixels exhibit a soft saturation and asymptotically reach

M. Dandin et al., *Single-Photon Avalanche Diodes and Photon Counting Systems*,
https://doi.org/10.1007/978-3-031-64334-7_4

their dead-time limited maximum count, which occurs at fluxes that are higher than those at which full-well capacity is reached in APS structures. Therefore, SPADs inherently possess a nonlinear signal compression feature at the front end that extends their dynamic range [10].

Despite these advantages, SPADs are far from ideal. They can have an excessive dark count rate (DCR) at or near room temperature, which can limit their performance in a passive imaging mode. During a given measurement period, they can produce a number of false counts resulting from carrier generation processes that occur even when there are no photons impinging on the device. For instance, the DCR of a SPAD can range from a few tens of Hz to over 1 MHz, depending on the device's architecture [14–20]. These *dark* counts are largely the result of thermal generation, afterpulsing, and band-to-band tunneling [21], and they set the device's noise floor (see Fig. 4.1).

These DCR components can be curtailed using several methods. To reduce the component that is due to thermal generation, the device may be cooled [22]. To limit afterpulsing, which consists of dark counts originating from the release of trapped carriers associated with past avalanches, the device may be held in a quenched state for a time period that is longer than the trapped carrier lifetime. Alternatively, the quenching time may be minimized in order to limit the number of carriers crossing the junction, which reduces the afterpulsing probability. We note that cooling may also affect the probability of afterpulsing because carrier lifetimes are temperature-dependent [23].

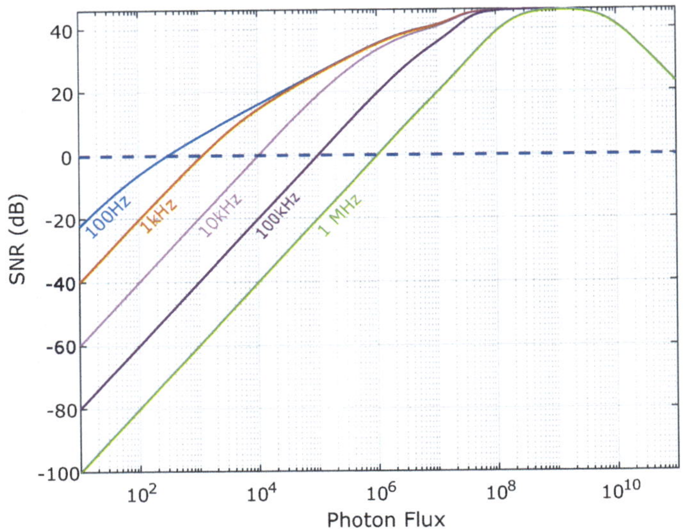

Fig. 4.1 Signal-to-noise ratio of a SPAD according to a theoretical model published by Ingle et al. [10]. From left to right the model is evaluated for a SPAD having dark count rates of 100 Hz, 1 kHz, 10k, 100 kHz, and 1 MHz. The point at which a trace crosses the dashed line (i.e., at SNR = 0 dB) is the noise-equivalent photon flux, and it sets the lower bound of the device's dynamic range

To reduce band-to-band tunneling, the semiconductor structure forming the SPAD may be designed such that the strength of the avalanche-causing field is lowered to minimize field-assisted carrier generation [24]. We note that this latter option is not available to a designer using a commercial CMOS process because he or she has no control over the fabrication process [25].

We introduced perimeter gating to CMOS SPAD imagers since it offers a technology agnostic means to reduce the DCR, and it is much simpler to implement than the abovenoted typical DCR mitigation strategies. Therefore, the goal of this chapter is to explore the use of perimeter gating in a passive imaging context. While there have been significant efforts (by the Dandin group and by others) towards studying the viability of perimeter gating as a dark count rate reduction method, its application to imaging has not yet been evaluated.

The work of Berkovich et al. and that of Habib et al. showed arrays of individually addressable perimeter-gated SPADs (pg-SPADs), a necessary feature for imaging, and they are noteworthy [26, 27]. However, these investigations did not fully explore the benefits of perimeter gating to imaging. For instance, Berkovich et al. showed a 20 × 20 array of pg-SPADs interfaced with an address-event representation (AER) readout framework for readout, but the effects of perimeter gating on array performance and on the AER readout scheme itself were not studied, nor was an image acquired with the device [26].

Similarly, Habib et al. studied the dark count rate dynamics of a small-density array of pg-SPADs (3 × 3), but this device could not be used to render an image due to its small size, and no discussions of the device's potential scalability or its application to imaging were provided [27]. Consequently, in this work, we seek to bridge this gap by demonstrating the use of perimeter gating in a pg-SPAD pixel array that is specifically configured for imaging.

Our work features a prototype imager made of a 64 × 64 array of p+/n-well pg-SPAD pixels interfaced with monolithically integrated active quenching and active reset (AQAR) circuits. The imager was fabricated in a three-metal, two-poly, twin-well, and low-voltage 0.35-μm CMOS process. Our investigation illustrates the potential of perimeter gating in the development of SPAD imagers that can operate at or near the quantum limit at room temperature. We have also explored several methods to compensate for the fixed pattern noise arising from the non-ideality of the imager and applied them on a test scene. Several images were captured at different voltages on the perimeter gate and analyzed both qualitatively and quantitatively.

4.2 Imager Architecture and Operation

Figure 4.2 illustrates several views of the fabricated prototype imager. Figure 4.2a shows the imager chip mounted directly on a printed circuit board (PCB). The chip's I/O pads were wirebonded directly to the PCB's pads, and the wirebonds were subsequently encapsulated using a double-dam fill epoxy process. The encapsulation

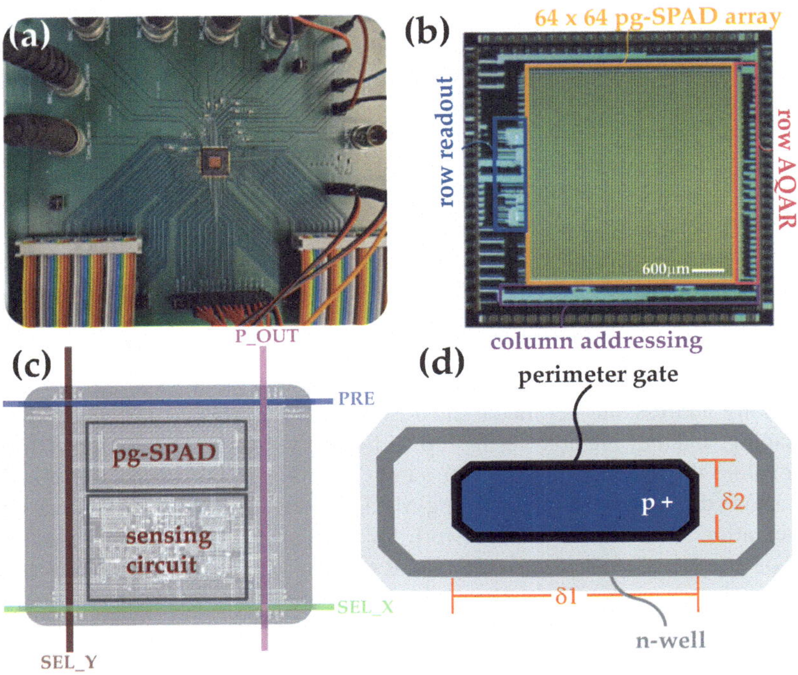

Fig. 4.2 (**a**) Photograph of the prototype pg-SPAD imager mounted and wirebonded to a printed circuit board. The wirebonds are encapsulated with a nonconductive epoxy. (**b**) Photomicrograph of the prototype and of some of its main constitutive circuit blocks, e.g., the pixel array, row readout circuitry, column addressing circuitry, and row-wise active quenching and active reset circuitry. (**c**) Layout view of a single pixel showing the pg-SPAD and its in-pixel sensing circuitry, along with address (X, Y) lines (SEL_X, SEL_Y), its output lines (P_OUT), and its hard reset (PRE) line. (**d**) Schematic of p+/n-well pixel, with dimensions $\delta_1 = 25\ \mu$m and $\delta_2 = 10\ \mu$m

step was undertaken in order to maintain the structural integrity of the wirebonds and to protect them from inadvertently being shorted during the handling of the board.

Figure 4.2b shows a photomicrograph of the chip taken with a bright-field upright optical microscope. The peripheral circuits include row addressing and row readout circuits, column addressing circuits, and one active quenching active reset (AQAR) circuit for each of the 64 rows. The pixel array includes 4096 pixels disposed in a 64×64 two-dimensional array. Figure 4.2c shows a layout view of a single pixel. Each pixel has its own in-pixel sensing circuitry, and it is surrounded by metal lines for X and Y addressing, for precharging the pixel with a hard reset pulse (PRE), and for outputting a digital signal corresponding to an avalanche event (P_OUT). Quenching and reset signals are provided to the pixel via a return bus upon the row-wise AQAR circuit receiving the pixel's output signal.

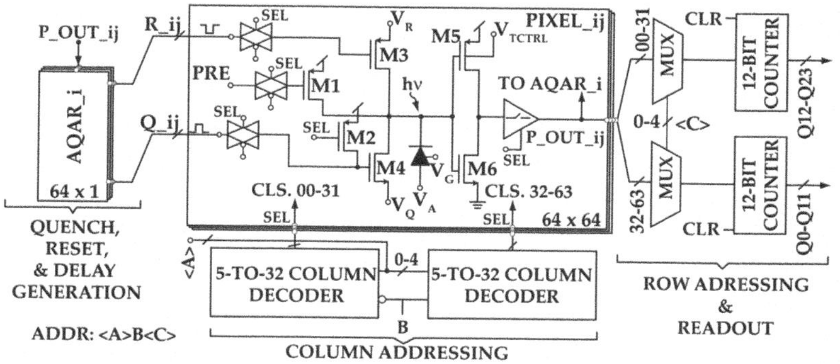

Fig. 4.3 Imager architecture. Each pixel includes selection and precharging circuitry, active quenching and active reset transistors, a pg-SPAD, and thresholding and output buffer circuits. Active quenching and active reset (AQAR) signals and quenching delay are generated row-wise

The pg-SPAD pixel is a p+/n-well junction with the perimeter gate implemented as a polysilicon layer that straddles the p+ anode. The perimeter gate and the anode are placed in an n-well cathode equipped with a peripheral contact ring. During operation, the n-well/p-sub parasitic diode is reverse biased as is typically done for integrated circuits fabricated in single-well CMOS processes. The pg-SPAD has an elongated shape ($\delta_1 = 25$ μm and $\delta_2 = 10$ μm) as it has been reported that defect centers that act as thermal generation centers are not uniformly distributed over an active area, and elongated diode shapes such as the one shown in Fig. 4.2d have a greater gettering efficiency.

Figure 4.3 shows a detailed schematic of the pixel sensing circuitry. Once pixels are selected by the **SEL** signals generated from the 5-to-32 column decoders, they are precharged through **M1**. Pixel integration initiates once the **PRE** signal is withdrawn. The cathode of an active diode starts to discharge through itself (to V_A) passively once a photon (**h**$\boldsymbol{\nu}$) is absorbed or a dark carrier triggers a false avalanche. Once the gate voltage of the inverter (**M5**, **M6**) crosses the threshold, its output toggles and is buffered to **P_OUT_ij** bus. Two multiplexers route the output signal to two 12-bit counters for being integrated throughout the integration window.

At the same time, **P_OUT_ij** triggers the AQAR block and initiate the active quench (**Q_ij**) and the subsequent active reset (**R_ij**) signal generation process. These signals discharge the cathode of the photodiode to $\mathbf{V_Q}$ through **M4** and recharge it back to $\mathbf{V_R}$ through **M3**. The hold time can be controlled by the delay control mechanism embedded in the AQAR block and the threshold of the sensing inverter can be controlled by $\mathbf{V_{TCTRL}}$. Pixels for readout are selected by asserting an 11-bit address $< \mathbf{A} > \mathbf{B} < \mathbf{C} >$ on the peripheral logic unit. Final counts $\mathbf{Q0} - \mathbf{Q11}$ and $\mathbf{Q12} - \mathbf{Q23}$ are read using an off-chip data acquisition module.

4.3 Theory and Background

4.3.1 Image Formation

In the simplest terms, a captured image is a 2-D matrix with each element representing the response from a specific pixel. This one-to-one correspondence between a scene element and a pixel (picture element) is established by the lens in the camera setup, making each pixel response ideally independent from that of other pixels. For a pg-SPAD imager in the photon-counting mode, this response is essentially the number of events (photon incidents or dark events) taking place within a fixed integration time. Hence, each pixel response, $R_i(n, V_G)$ at the ith location under an average incoming photon arrival rate, n, while the pg-SPAD is operating with a perimeter gate voltage, V_G, can be modelled as follows [28]:

$$R_i(n, V_G) = P_i \cdot (PRNU_i(V_G) \cdot I_i(n) + DSNU_i(V_G)) \tag{4.1}$$

Here, P_i is the read pattern factor stemming from the readout mechanism and is assumed to be independent of the incoming number of photons or the applied perimeter gate voltage. $I_i(n)$ represents the ideal image pixel to be recovered which is scaled by the photoresponse nonuniformity (PRNU) and added to the dark signal nonuniformity (DSNU). Both PRNU and DSNU are assumed to be affected by the perimeter gate voltage since perimeter gating physically alters the effective active area by modulating the perimeter junction electric field. The dot-product, "." represents element-wise matrix multiplication. For simplicity, we will omit the subscript for the coordinate (i.e., i) in the subsequent equations and all mathematical operators will imply element-wise operations unless stated otherwise.

4.3.2 Elimination of DSNU with Perimeter Gating

From Eq. 4.1, we can think of at least two ways to eliminate DSNU from the pixel response. In one method, we could acquire the response in the dark, $R_{dark}(n = 0, V_G)$ (as defined in Eq. 4.2), and subtract it from Eq. 4.1:

$$R_{dark}(n = 0, V_G) = P \cdot DSNU(V_G); \; since \; I(n) = 0 \tag{4.2}$$

In another method, we could operate the imager at higher perimeter gate regime with vanishing dark count probability [29]. Figure 4.4 shows the sensor response in the dark when the perimeter gate voltage magnitude is varied from 0 V to 8 V. The integration time was 10 ms. It is interesting to note the gradual disappearance of the dark noise i.e., in the images for 4 and 5 V, and its complete elimination from 6 V and above.

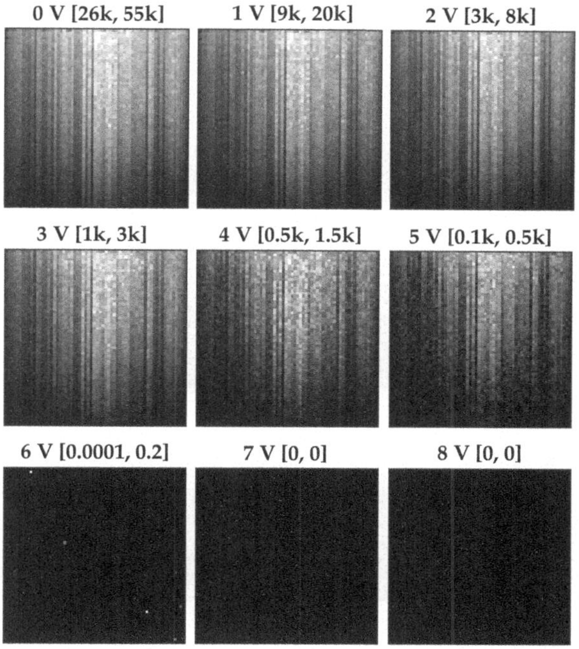

Fig. 4.4 Sensor response in the dark: average dark count rate and maximum dark count rate, both gradually decreased with the increase of the perimeter gate voltage magnitude and eventually vanished at high perimeter gate voltages (values are displayed with each subplot as V_G [avg. count rate, max. count rate]). 10 ms was used as the integration time. All images are displayed with normalized colormap, [0 1]

4.3.3 Effects on PRNU from Perimeter Gating

Both DSNU elimination methods will generate an output response, $O(n, V_G)$, that is linearly proportional to the number of incident photons until dead-time-induced nonlinearity dominates or the photon counter saturates under uniform illumination. This response is modelled as follows:

$$O(n, V_G) \approx P \cdot PRNU(V_G) \cdot I(n) \tag{4.3}$$

Although Eq. 4.3 is free from the random offset contributed by the dark counts, the slope of the response curve still varies spatially due to pixel-to-pixel $PRNU$. As a result, we see an FPN pattern instead of a uniform image when the sensor is illuminated homogeneously as shown in Fig. 4.5.

The top panel in Fig. 4.5 shows the photoresponse at low perimeter gating before ($R(n, V_G)$) and after ($O(n, V_G)$) dark response subtraction, while the bottom panel displays dark signal free photoresponse at high perimeter gating. We can notice the emergence of a distinctive pattern as the dark noise is eliminated, especially at higher perimeter gate voltages.

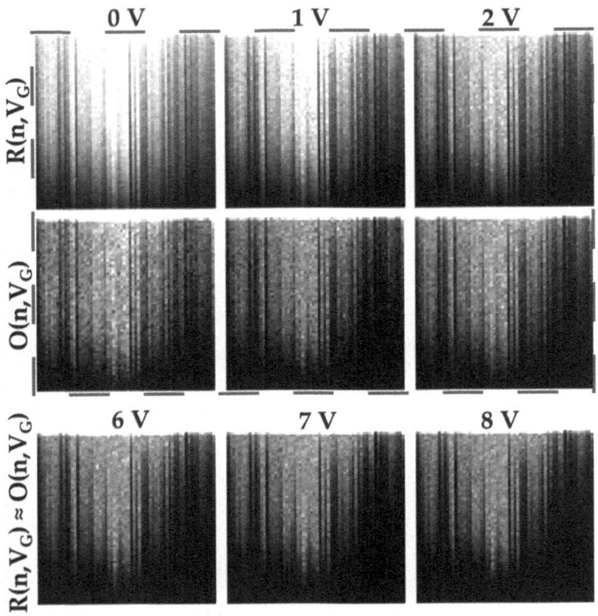

Fig. 4.5 FPN dominated by the read pattern factor: (top panel) photoresponse at low perimeter gating before ($R(n, V_G)$) and after ($O(n, V_G)$) dark response subtraction. (bottom panel) Dark signal free photoresponse at high perimeter gating. Sensor was uniformly illuminated directly for an integration time of 10 ms. Gate voltages are displayed on top of the images

4.3.4 Fixed Pattern Noise Compensation

FPN correction methods are categorized into two broad groups, namely, calibration-based and scene-based methods. Calibration-based methods include the two-point method, the subsection correction method, the S-curve method, and the polynomial fitting method.

However, all these methods rely on the response curves of *all* the pixels for a wide range of incoming flux depending on the pixels' full-well capacity. Unfortunately, these response curves vary substantially with changes in operating conditions, i.e., temperature, applied bias voltage, and integration time. This makes these methods less appropriate and impractical in a dynamic environment.

Furthermore, with these methods, recalibration is required periodically due to sensor aging. Nevertheless, systems with low computational power can benefit from the simple two-point correction method especially when the sensor response is appreciably linear.

On the other hand, alternative scene-based methods include filter-based, statistics-based, and optimization-based methods. Here, we show several methods for FPN correction. Specifically, we present a two-point calibration method, a wavelet filtering method, and a multichannel least-mean square-based optimization method.

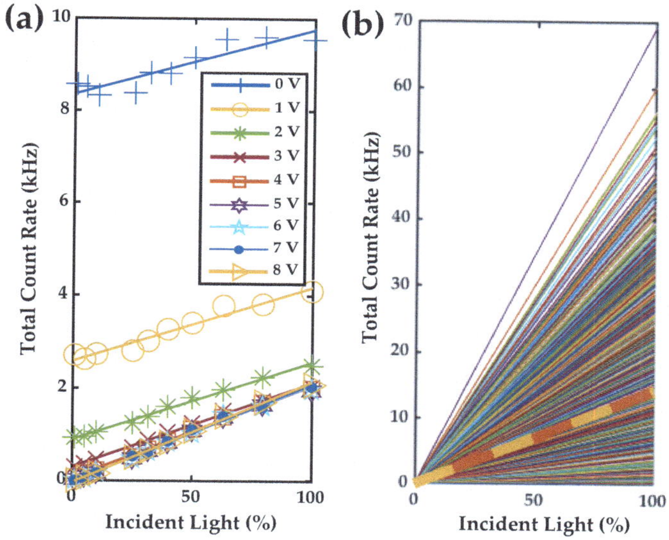

Fig. 4.6 Photoresponse variation with perimeter gating: (**a**) offset elimination and linearity improvement for a random pixel and (**b**) response variation across the array for 4096 pixels (thick line represents the average response)

4.3.4.1 Two-Point Calibration

Figure 4.6a shows the response of a randomly selected pixel at different perimeter gate voltages while the incident photon rate was reduced from 100% to 0% using a motorized neutral density filter wheel. The solid lines depict the fitted straight-line through the total counts.

As expected from Eq. 4.1, the response is linear with respect to the incoming photon rate. With increasing perimeter gate voltage, we notice two changes. First, the offset due to the dark signal gradually vanishes and the lines pass through the origin and they coincide (following Eq. 4.3) at higher perimeter gate voltages (> 5 V). Second, the "goodness of fit" for the fitted curves improves from 0.841 to 0.998, implying improved linearity in the response resulting from DSNU suppression. Hence, we plot the response curves (fitted lines) for all the pixels in Fig. 4.6b with 8 V on the gate. Unsurprisingly, the response curves display variation in slopes owing to the fact that the slope $m(V_G) = P \cdot PRNU(V_G)$ varies pixel-wise resulting in the FPN in Fig. 4.5.

In the two-point calibration method, a single (average) response curve is formulated as shown below:

$$R_{avg}(V_G) = m_{avg}(V_G) \cdot I(n) + o_{avg}(V_G) \tag{4.4}$$

Here, $m_{avg}(V_G) = \sum P_i \cdot PRNU_i(V_G)$ and $o_{avg}(V_G) = \sum P_i \cdot DSNU_i(V_G)$ are the average slope and offset from all the curves, respectively (the thick line in Fig. 4.6b).

Once the average response curve is obtained, the following equation transforms the acquired image, I_{acq}, into the corrected image, I_{cor}, as shown below.

$$I_{cor} = (\frac{I_{acq} - o}{m}) \cdot m_{avg} + o_{avg} \tag{4.5}$$

Here, m and o are the slope and the offset for individual pixels.

4.3.4.2 Wavelet Decomposition

While filter-based FPN removal leverages the periodicity in the pattern (specifically column FPN), any image component of the same spatial frequency will also be affected by it. Therefore, we resorted to read pattern factor, P, and $PRNU$ estimation from the vectorized slope variation shown in Fig. 4.7a. Using second-

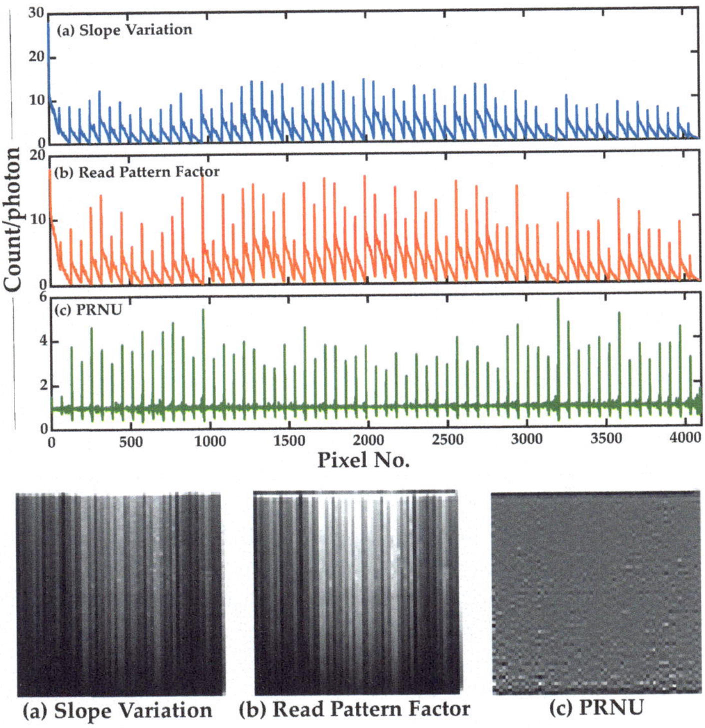

Fig. 4.7 Wavelet decomposition of the slope variation: (top panel) vectorized form and (bottom panel) 2-D image of (**a**) the slope variation, (**b**) read pattern factor, and (**c**) PRNU. Second-order *symlets* were used as the mother wavelet for the decomposition of the vectorized slope variation into (**a**) low-frequency approximation (**b**) and high-frequency detailed (**c**) components

order *symlets* as the mother wavelet because of their closeness to the signal of interest, we decomposed the logarithm of the slope variation into the low-frequency approximation component (labelled as the read pattern factor) and the high-frequency detailed component (labelled as the PRNU) as outlined below:

$$log(m) = log(P) + log(PRNU) \tag{4.6}$$

$$P = antilog(approximation(log(m))) \tag{4.7}$$

$$PRNU = antilog(detail(log(m))) \tag{4.8}$$

Here, $approximation(\cdot)$ and $detail(\cdot)$ represent the approximation and the detail signal components after a single-level decomposition.

There are at least two merits with this approach compared to the two-point calibration technique. First, we are not required to generate the response curve. Rather, we can apply a wavelet transform on a single photoresponse image taken at a higher perimeter gate voltage (free from DSNU). Second, the level of decomposition can be varied to filter out any illumination nonuniformity (if any), assuming the variation is slower compared to the read pattern factor. The extracted read pattern factor can be used to correct its dominant contribution from the FPN in the acquired images.

4.3.4.3 Multichannel Least Mean Square (mcLMS)

In this method, we can correct the FPN from the acquired images on the fly, without needing any pre-estimated read pattern factor or slope correction factor. As a result, recalibration is unnecessary. All that is required is a reference image of the FPN (ideally a scene as uniformly illuminated as possible).

Assuming I_{ref} is such an FPN reference image and I_{acq} is the FPN-corrupted acquired image, Fig. 4.8 outlines the mcLMS method for FPN correction.

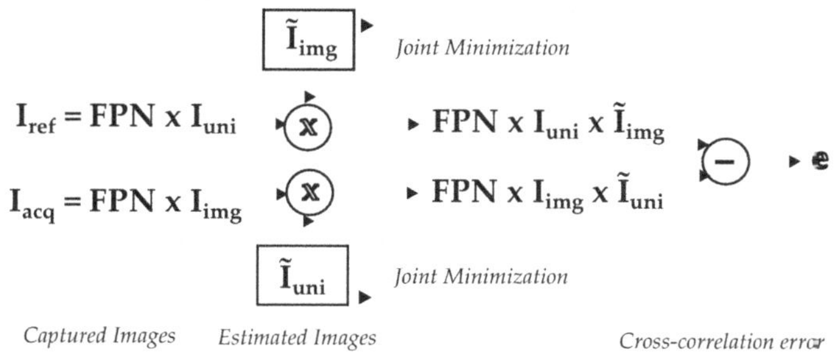

Fig. 4.8 Multichannel least-mean-square (mcLMS) algorithm for FPN correction. The captured images contain FPN as a common factor to be eliminated by joint estimation of the corresponding images while minimizing the cross-correlation error

The model features two data channels. For a reference image, i.e., the first channel, we obtained a uniform scene with the camera to be used as I_{ref}. Any other scene can be captured and considered as the other channel. Under this condition the acquired images can be written as the following:

$$I_{ref} = FPN \times I_{uni} \tag{4.9}$$

$$I_{acq} = FPN \times I_{img} \tag{4.10}$$

Using an estimate for each channel's true image (i.e., $\tilde{I}_{uni}$ and $\tilde{I}_{img}$), we can generate a cross-correlation error, $e = FPN \times (\tilde{I}_{uni} \cdot I_{img} - \tilde{I}_{img} \cdot I_{uni})$. Employing the adaptive-step least-mean-squared (LMS) algorithm to minimize the cross-correlation error energy, the estimates gradually converge to the true target images with increasing accuracy, although a plateau is reached. While the derivation of the cross-correlation error did not include any residual error from the DSNU, the mcLMS method still provides the best estimates for the true images depending on the DSNU's strength.

4.4 Results and Discussion

This section presents our results from different experiments conducted with this image sensor. We apply the theory discussed in the previous section to demonstrate the imager's performance as a function of perimeter gating.

4.4.1 Reduction of Nonuniformity with Perimeter Gating

Although Fig. 4.4 displays gradual disappearance of the array-wide dark counts in a normalized color map for qualitative evaluation, a single metric for quantification of DSNU was not presented there. Here, we quantify the DSNU as the array-wide variance of dark counts (inset of Fig. 4.9a) with perimeter gate voltage magnitude. Also, a histogram is plotted for the pixel-wise dark count deviation from the array average as we increased the perimeter gate voltage magnitude.

The DSNU becomes zero for gate voltage greater or equal to 3.5 V, earlier than what we have seen before, because we reduced the integration time from 10 ms to 100 μs. It was done to be consistent with typical imaging experiments, which were done at higher frame rate. Nevertheless, the trend of the DSNU histogram remains the same, i.e., the histogram gradually reduces to zero with increasing perimeter gate voltage magnitude.

Confirming that the PRNU is affected by perimeter gating, Fig. 4.9b shows a deceasing trend for the standard deviation of the PRNU defined in Eq. 4.8. Since the

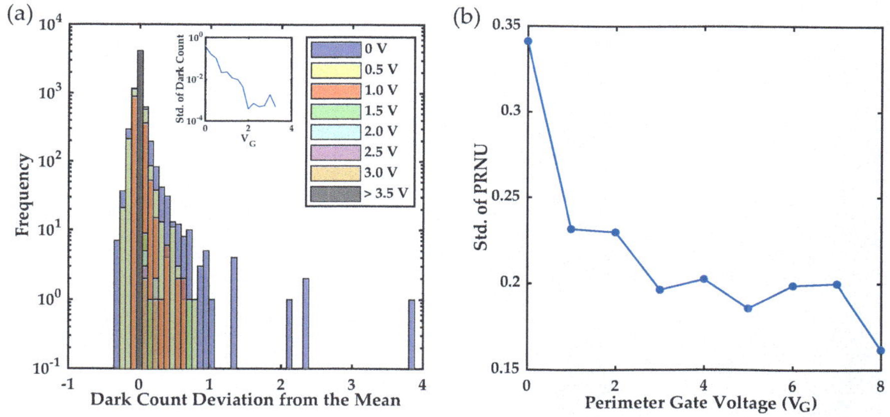

Fig. 4.9 DSNU and PRNU reduction with perimeter gating. (**a**) Histogram for the pixel-wise dark count deviation from the mean. Standard deviation of the array-wide dark counts is used to quantify DSNU (inset). (**b**) Reduction of PRNU as a function perimeter gate voltage. To reduce the effect of the photon shot noise, we use 10 ms as the integration time for this experiment

PRNU is also affected by the photon shot noise, we used 10 ms as the integration time for this experiment to mitigate the effect of the source noise. Although the change is not as dramatic as that of the DSNU, these results support our initial proposition that the PRNU is also modulated by the perimeter gate voltage V_G.

4.4.2 *Raw Image Acquisition at Varying Gate Voltages*

Figure 4.10a displays the captured raw images at different perimeter gate voltages. We can see the dominant FPN in each image. However as expected, the effect of DSNU is more prominent at lower perimeter gating (i.e., images in the first row).

To evaluate the effect of DSNU in the captured images, Fig. 4.10b shows the mean square error (MSE) of the difference image between adjacent gate voltage raw images. It can be easily related to the DSNU histogram as we see the MSE plateau near 4 V.

4.4.3 *FPN Correction*

Here, we present FPN-corrected images following the procedure outlined previously for (i) the two-point calibration method, (ii) the wavelet decomposed read pattern factor compensation technique, and (iii) the mcLMS optimization-based technique.

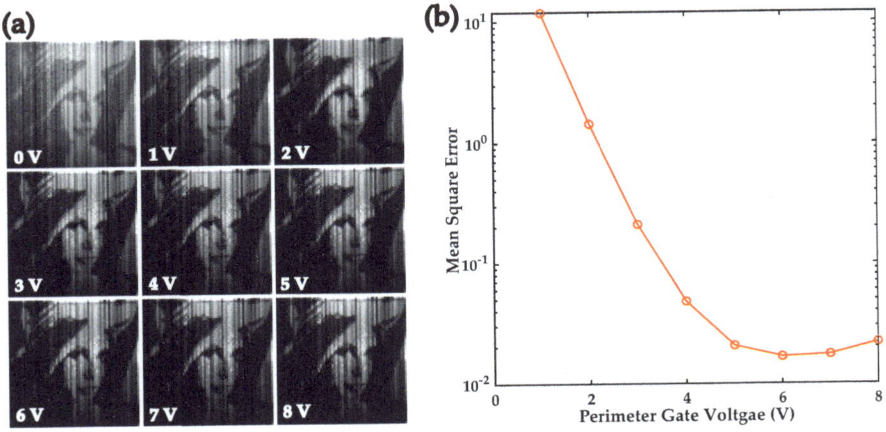

Fig. 4.10 Raw image evaluation at increasing perimeter gate voltage. (**a**) Captured images and (**b**) the mean square error plot between adjacent images show the effect of DSNU suppression. However, the residual PRNU becomes prominent at higher gate voltage that requires FPN compensation

4.4.3.1 Calibration and Filtered-Based Methods

The FPN-corrected images are displayed in Fig. 4.11a. Comparing the top and the bottom rows, we easily identify the benefit of DSNU suppression before applying the two-point scale correction method. Since offset estimation is not needed at high perimeter gate voltages, two-point calibration shows better FPN compensation in this regime. Although the middle row suffers from contrast variation, the images are still FPN-free. We did not perform any additional image processing, i.e., contrast enhancement, hot pixel correction, etc., to demonstrate the effect of this simple correction algorithm alone.

Figure 4.11b shows the images corrected using the read pattern factor compensation estimated using Eq. 4.7. Although we see a better contrast consistency compared to the two-point correction method, the top row images still bear the evidence of the effect of DSNU when it is not properly suppressed. That is, even dark frame subtraction did not give a better result than the DSNU suppression using high-voltage perimeter gating.

4.4.3.2 McLMS Correction

Finally, we present the FPN-corrected images through mcLMS optimization in Fig. 4.12. This time, we have also included the original image.

Visual comparison reveals that the corrected images are better in quality compared to the previous two methods, especially at low perimeter gating. However, the benefit of DSNU suppression is pronounced at high perimeter gate regime when we compare the images with the original.

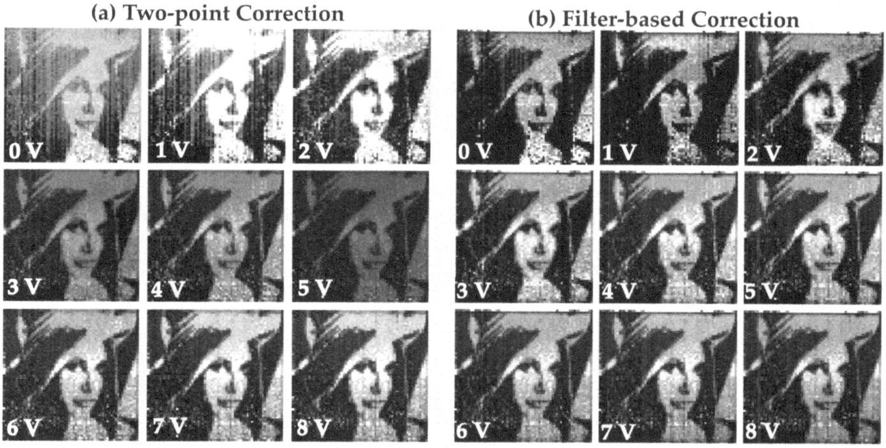

Fig. 4.11 FPN correction using calibration-based and filter-based correction method. (**a**) Two-point calibration suffers from contrast variation unlike the filtered-based (**b**) correction method

Fig. 4.12 FPN correction using mcLMS optimization

4.5 Conclusions

In this chapter, we have demonstrated the imaging capabilities of a 64 × 64 perimeter-gated single-photon avalanche diode (pg-SPAD) imager in the context of passive imaging. The results show that with the increase of the perimeter gate voltage magnitude, both DSNU and PRNU decreased, and for short integration times, the DSNU vanished at the higher perimeter gate regime while PRNU saturated. Furthermore, optimization-based read pattern correction resulted in better-quality images, significantly reducing pixel-to-pixel nonuniformity in photoresponse.

References

1. J. Kostamovaara, S. Jahromi, L. Hallman, G. Duan, J.-P. Jansson, P. Keranen, Solid-state pulsed time-of-flight 3-D range imaging using CMOS SPAD focal plane array receiver and block-based illumination techniques. IEEE Photon. J. **14**(2), 1–11 (2022). https://doi.org/10.1109/JPHOT.2022.3153487
2. C. Niclass, A. Rochas, P. Besse, E. Charbon, Design and characterization of a CMOS 3-D image sensor based on single photon avalanche diodes. IEEE J. Solid State Circuit **40**(9), 1847–1854 (2005)
3. M. Albota, R. Heinrichs, D. Kocher, D. Fouche, B. Player, M. O'Brien, B. Aull, J. Zayhowski, J. Mooney, B. Willard, R. Carlson, Three-dimensional imaging laser radar with a photon-counting avalanche photodiode array and microchip laser. Appl. Opt. **41**(36), 7671–7678 (2002)
4. R. Heinrichs, B. Aull, R. Marino, D. Fouche, A. McIntosh, J. Zayhowski, T. Stephens, M. O'Brien, M. Albota, Three-Dimensional Laser Radar with APD Arrays, in *Laser Radar Technology and Applications VI*, ser. Proceedings of the Society of Photo-Optical Instrumentation Engineers (SPIE), ed. by GW, Kamerman, vol. 4377, pp. 106–117 (2001), Conference on Laser Radar Technology and Applications VI, FL, APR 17–19, 2001
5. V. Zickus, M.L. Wu, K. Morimoto, V. Kapitany, A. Fatima, A. Turpin, R. Insall, J. Whitelaw, L. Machesky, C. Bruschini, D. Faccio, E. Charbon, Fluorescence lifetime imaging with a megapixel SPAD camera and neural network lifetime estimation. Sci. Rep. **10**(1), 1–10 (2020). https://doi.org/10.1038/s41598-020-77737-0
6. D.E. Schwartz, E. Charbon, K.L. Shepard, A single-photon avalanche diode array for fluorescence lifetime imaging microscopy. IEEE J. Solid State Circuits **43**(11), 2546–2557 (2008). https://doi.org/10.1109/JSSC.2008.2005818
7. S.W. Hutchings, N. Johnston, I. Gyongy, T. Al Abbas, N.A. Dutton, M. Tyler, S. Chan, J. Leach, R.K. Henderson, A reconfigurable 3-D-stacked SPAD imager with in-pixel histogramming for flash LIDAR or high-speed time-of-flight imaging. IEEE J. Solid State Circuits **54**(11), 2947–2956 (2019). https://doi.org/10.1109/JSSC.2019.2939083
8. A. Katz, T. Blank, A. Fenigstein, T. Leitner, Y. Nemirovsky, Active-reset for the N+P single-ended SPAD used in the NIR LiDAR receivers. IEEE Trans. Electron Dev. **66**(12), 5191–5195 (2019). https://doi.org/10.1109/TED.2019.2946015
9. S. Ma, S. Gupta, A.C. Ulku, C. Bruschini, E. Charbon, M. Gupta, *Quanta Burst Photography* (2020). arXiv e-prints. arXiv:2006.11840. https://doi.org/10.48550/arXiv.2006.11840
10. A. Ingle, A. Velten, M. Gupta, High Flux Passive Imaging with Single-Photon Sensors, in *Proceedings of the IEEE Computer Society Conference on Computer Vision and Pattern Recognition*, vol. 2019 (IEEE Computer Society, 2019), pp. 6753–6762. https://doi.org/10.1109/CVPR.2019.00692. ISBN:9781728132938. ISSN:10636919
11. A. Ingle, T. Seets, M. Buttafava, S. Gupta, A. Tosi, M. Gupta, A. Velten, Passive Inter-Photon Imaging, in *2021 IEEE/CVF Conference on Computer Vision and Pattern Recognition (CVPR)* (IEEE, 2021), pp. 8581–8591. https://doi.org/10.1109/CVPR46437.2021.00848. ISBN:978-1-6654-4509-2
12. T. Merhav, V. Savuskan, Y. Nemirovsky, Gun Muzzle Flash Detection Using CMOS Sensors, in *2013 IEEE International Conference on Microwaves, Communications, Antennas and Electronic Systems (COMCAS 2013)* (IEEE, 2013), pp. 1–4. https://doi.org/10.1109/COMCAS.2013.6685317. ISBN 978-1-4673-5756-2
13. A. Katz, A. Shoham, C. Vainstein, Y. Birk, T. Leitner, A. Fenigstein, Y. Nemirovsky, Passive CMOS single photon avalanche diode imager for a gun muzzle flash detection system. IEEE Sens. J. **19**(14), 5851–5858 (2019). https://doi.org/10.1109/JSEN.2019.2904312
14. H. Finkelstein, M. Hsu, S. Esener, STI-bounded single-photon avalanche diode in a deep-submicrometer CMOS technology. IEEE Electron Dev. Lett. **27**(11), 887–889 (2006). https://doi.org/10.1109/LED.2006.883560

15. D. Palubiak, M.M. El-Desouki, O. Marinov, M.J. Deen, Q. Fang, High-speed, single-photon avalanche-photodiode imager for biomedical applications. IEEE Sens. J. **11**(10), 2401–2412 (2011). https://doi.org/10.1109/JSEN.2011.2123090
16. J.A. Richardson, E.A.G. Webster, L.A. Grant, R.K. Henderson, Scaleable single-photon avalanche diode structures in nanometer CMOS technology. IEEE Trans. Electron Dev. **58**(7), 2028–2035 (2011). https://doi.org/10.1109/TED.2011.2141138
17. M. Gersbach, C. Niclass, E. Charbon, J. Richardson, R. Henderson, L. Grant, A Single Photon Detector Implemented in a 130nm CMOS Imaging Process, in *ESSDERC 2008 - 38th European Solid-State Device Research Conference* (IEEE, 2008), pp. 270–273. https://doi.org/10.1109/ESSDERC.2008.4681750. ISBN:978-1-4244-2363-7
18. M. Marwick, A. Andreou, Single photon avalanche photodetector with integrated quenching fabricated in TSMC 0.18[micro sign]m 1.8V CMOS process. Electron. Lett. **44**(10), 643 (2008). https://doi.org/10.1049/el:20080673
19. C. Niclass, C. Favi, T. Kluter, M. Gersbach, E. Charbon, A 128 ×128 single-photon image sensor with column-level 10-Bit time-to-digital converter array. IEEE J. Solid State Circuits **43**(12), 2977–2989 (2008). https://doi.org/10.1109/JSSC.2008.2006445
20. C. Niclass, C. Favi, T. Kluter, M. Gersbach, E. Charbon, A 128×128 Single-Photon Imager with On-Chip Column-Level 10b Time-to-Digital Converter Array Capable of 97ps Resolution, in *2008 IEEE International Solid-State Circuits Conference - Digest of Technical Papers* (IEEE, 2008), pp. 44–594. https://doi.org/10.1109/ISSCC.2008.4523048. ISBN:978-1-4244-2010-0
21. M. Ghioni, A. Gulinatti, I. Rech, F. Zappa, S. Cova, Progress in silicon single-photon avalanche diodes. IEEE J. Sel. Topics Quantum Electron. **13**(4), 852–862 (2007). https://doi.org/10.1109/JSTQE.2007.902088
22. B. Patra, R.M. Incandela, J.P.G. van Dijk, H.A.R. Homulle, L. Song, M. Shahmohammadi, R.B. Staszewski, A. Vladimirescu, M. Babaie, F. Sebastiano, E. Charbon, Cryo-CMOS circuits and systems for quantum computing applications. IEEE J. Solid State Circuits **53**(1), 309–321 (2018). https://doi.org/10.1109/JSSC.2017.2737549
23. B. Korzh, H. Zbinden, *Low Temperature Performance of Free-Running InGaAs/InP single-Photon Negative Feedback Avalanche Diodes*, ed. by M.A. Itzler, J.C. Campbell (2014), p. 91140O. https://doi.org/10.1117/12.2050487
24. T.C. de Albuquerque, D. Issartel, R. Clerc, P. Pittet, R. Cellier, F. Calmon, Lowering the Dark Count Rate of SPAD Implemented in CMOS FDSOI Technology, in *2019 Joint International EUROSOI Workshop and International Conference on Ultimate Integration on Silicon (EUROSOI-ULIS)* (IEEE, 2019), pp. 1–4. https://doi.org/10.1109/EUROSOI-ULIS45800.2019.9041916. ISBN:978-1-7281-1658-7
25. C. Niclass, M. Gersbach, R. Henderson, L. Grant, E. Charbon, *A 130-nm CMOS Single-Photon Avalanche Diode*, ed. by J. Piprek, J. . Wang (2007), p. 676606. https://doi.org/10.1117/12.728878
26. A. Berkovich, T. Datta-Chaudhuri, P. Abshire, A Scalable 20 x 20 Fully Asynchronous SPAD-Based Imaging Sensor with AER Readout, in *2015 IEEE International Symposium on Circuits and Systems (ISCAS)* (IEEE, 2015), pp. 1110–1113. https://doi.org/10.1109/ISCAS.2015.7168832. ISBN:978-1-4799-8391-9
27. M.H.U. Habib, M.S.A. Shawkat, N. McFarlane, A Tunable Single Photon Avalanche Diode Pixel with Improved Time Resolution, in *2017 IEEE SENSORS* (IEEE, 2017), pp. 1–3. https://doi.org/10.1109/ICSENS.2017.8234050. ISBN:978-1-5090-1012-7
28. T. Zhang, X. Li, J. Li, Z. Xu, CMOS fixed pattern noise removal based on low rank sparse variational method. Appl. Sci. **10**(11), 3694 (2020). https://doi.org/10.3390/app10113694
29. M.S. Sajal, K.-C. Lin, B. Senevirathna, S. Lu, M. Dandin, Perimeter-Gated Single-Photon Avalanche Diode Imager with Vanishing Room Temperature Dark Count Probability, in *2022 29th IEEE International Conference on Electronics, Circuits and Systems (ICECS)* (IEEE, 2022), pp. 1–4. https://doi.org/10.1109/ICECS202256217.2022.9970824. ISBN:978-1-6654-8823-5

Chapter 5
Perimeter-Gated Single-Photon Avalanche Diode Arrays as Hardware Security Primitives

5.1 Introduction

Physically unclonable functions (PUFs) rely on inherent silicon manufacturing process variations that impart unique physical attributes from chip to chip [1–3]. By leveraging these attributes, chip-specific unique signatures can be extracted. Interestingly, these signatures cannot be predicted and they cannot be reproduced even if the signatures from some other chips of the same batch are available. They can thus be used for generating authentication keys that are chip-specific [4, 5].

Traditionally, authentication keys are stored within a device and retrieved prior to the device being authenticated [6]. However, with the explosive growth of interconnected devices and the Internet of Things (IoT), peripheral hardware have become vulnerable to tampering, particularly to attempts of impersonating trusted devices [7]. As such, PUFs help by eliminating the need to store keys physically on the device itself. Rather, leveraging the device's PUF, authentication keys can be generated on the fly in response to a known challenge, thus creating what is known as a challenge-response pair (CRP) that can be used for authentication [8]. The size of the CRP space determines the PUF's strength [9] and ensures the longevity of the PUF source.

Figure 5.1 shows a generic authentication and key distribution protocol for secured communication. Once a ***Client*** requests the ***Key Distribution Center*** (KDC) for a session key to establish a connection with a ***Server***, the KDC wraps a session key with the client's long-term key upon verification. Usually these long-term keys are derived from the client's password. Afterwards, a mutual authentication protocol is followed by the client and the server to establish a secured session [10].

These keys can be generated from the client's PUF, guaranteeing their uniqueness each time KDC challenges them with the stored CRPs in its secured database. As a result, an impersonator cannot copy the keys since the client's responses are never stored locally. Hence, the benefit of PUF-based protocols enhance the security in

M. Dandin et al., *Single-Photon Avalanche Diodes and Photon Counting Systems*,
https://doi.org/10.1007/978-3-031-64334-7_5

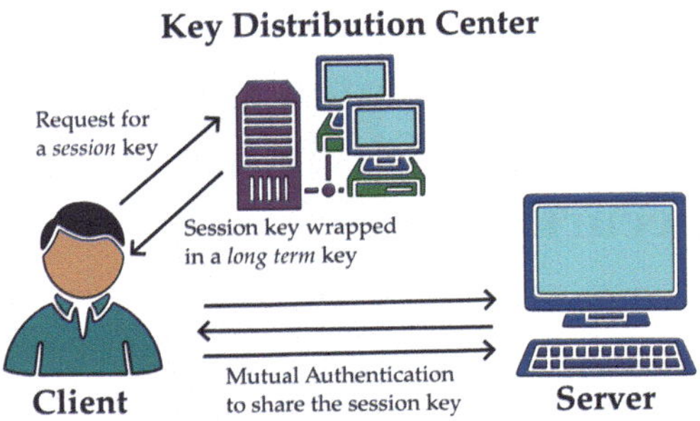

Fig. 5.1 Needham-Schroeder protocol for secured key distribution. These keys can be generated from the client's PUF for client verification

such networked systems. Example implementations of silicon PUFs include arbiter chains [11, 12], ring oscillators, [13, 14] and memory circuits, [15, 16]. In each case, inherent manufacturing process variations impart a key signature to these circuits, making them suitable for generating unique and device-specific keys.

In recent years, CMOS image sensor chips have become ubiquitous, and they too have been studied as a means of generating authentication keys from PUFs. For example, Cao et al. exploited the reset voltage variation from pixel pairs of a CMOS image sensor to generate a unique and reliable binary signature [17]. Kim et al. showed a PUF based on an imager's column fixed-pattern noise (CFPN) [18]. Valsesia et al. implemented an imager PUF based on photoresponse nonuniformity (PRNU) [19]. Furthermore, Kim et al. used the high-frequency components of an APS imager's dark signal nonuniformity (DSNU) to generate PUF-based keys [20].

However, these notable approaches are not readily applicable to cameras that operate in Geiger mode, i.e., in cameras configured for high-speed imaging based on light quanta and for photon counting in general [21, 22]. As these cameras are becoming prominent in autonomous navigation and in other emerging technological applications [23, 24], we offer a strategy for developing PUFs and authentication keys from them. Our test bed is a custom-designed Geiger-mode camera implemented from an array of perimeter-gated single-photon avalanche diodes (pg-SPADs) [25–28] which has been introduced in Chap. 4.

We characterized three of these 64×64 pg-SPAD imagers, fabricated in $0.35-\mu m$ standard CMOS process [29–31], over a wide temperature range (27 °C to 75 °C), and we show that the imagers' dark count maps (DCMs) may be used as a PUF from which unique authentication keys can be generated. We show that the keys generated from the DCMs have high inter-chip normalized Hamming distance (nHD) for the same challenge, suggesting that unique keys can be generated from each sensor [32].

We further show that the intra-chip nHD between different keys from different challenges is sizable, thus allowing a large number of distinct CRPs to be generated from the same chip. For instance, the number of distinct m-bit challenges from a sensor with N pixels will be P_m^N, where P is the number of possible permutations. Furthermore, we show that the responses are resilient against changes in temperature. Lastly, and most importantly, we show that the ability of perimeter gating to modulate the dark count rate (DCR) at the single pixel level serves to change or conceal the imager's PUF, thereby providing an additional security feature [1].

We also demonstrate a novel technique for increasing the CRP space of the aforementioned pg-SPAD-based PUF [2]. Specifically, we show that perimeter gating may be used to dynamically alter the PUF and thus increase the number of unique challenge-response pairs. We show that this technique can overcome the typical size limitations of the CRP space of conventional CMOS imager-based PUFs [33]. We also explore pg-SPADs as true random number generators [3].

5.2 Imager Operation and Key Generation

5.2.1 Imager Operation in the PUF Mode

As previously noted, our test bed consisted of a CMOS imager having pg-SPADs as its constitutive pixels. The device's detailed architecture and its operation as an imager have been described in Chap. 4. Here, we only discuss the imager's operation in the PUF mode. The imager's individual pixels (**PUF unit**) may be selected and read out via a random access architecture implemented using column and row selection circuits. A timing diagram of the device's operation is shown in Fig. 5.2. In the context of devising CRPs, the imager is operated in the dark, and a given pixel address represents a single *challenge*, **C**, which returns a single *response*, **R**, consisting of the number of dark counts that occurred at the selected pixel.

Figure 5.3a shows the DCR of the pixel exhibiting the maximum DCR, that of the pixel exhibiting the minimum DCR, and the array average DCR of all the 4,096 pixels. The data confirm that there can be a sizable increase in dark count rates across the array as a function of temperature which typically follows an exponential trend [34].

Figure 5.3b shows the responses collected from the entire array of the three chips we tested. The responses form a DCM consisting of the room temperature dark counts collected at each pixel during the measurement time. The DCMs obtained with this procedure are normalized and plotted with a color scheme that highlights the differences in value at each pixel and for each chip. The collected DCMs thus form the imagers' PUFs, and as expected, the DCMs are different for each chip.

In order to be the basis of a robust authentication scheme, an imager's PUF must not drift significantly with temperature. We thus studied how temperature changes affected the DCMs, especially considering that SPADs can exhibit large increases in dark counts at elevated temperatures. This is because a significant portion of a SPAD's dark count rate is due to thermally generated carriers.

To ascertain at which temperature the normalized DCMs start to deviate significantly from their room temperature counterparts, we computed the normalized DCMs' mean square error (MSE) across the temperature range; specifically, the MSE measures how similar two normalized DCMs are from one temperature point

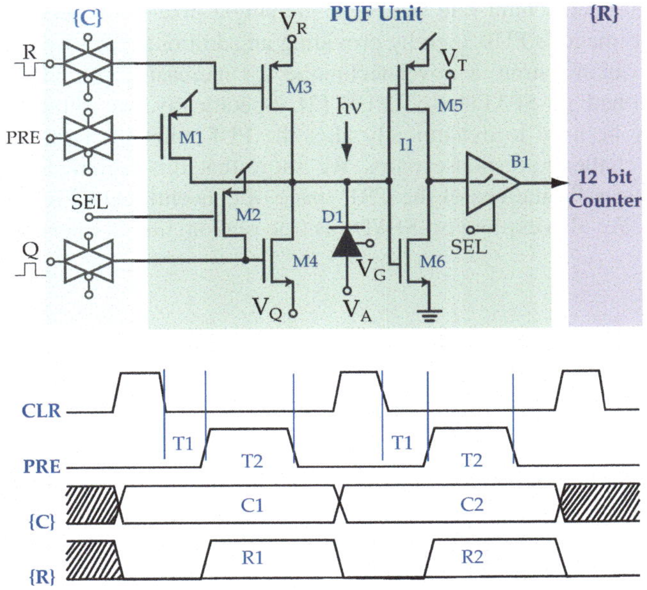

Fig. 5.2 Timing diagram of the pg-SPAD imager and challenge-response pair formation with reference to the pixel architecture described in Chap. 4. The output counter is first cleared (CLR) and a pixel address challenge, C1, is asserted; this address is from the challenge set $\{C\}$. The selected pixel is then precharged with a hard reset signal (PRE) in phase T1, and when this signal is de-asserted in phase T2, the pixel is readout by the output counter, providing a dark count value R1, which forms the response set $\{R\}$. This cycle is repeated to collect a full set of single challenge-response pairs

to another (see Fig. 5.3c). The data show that significant increases in MSE occur at and above 55 °C, suggesting that the imagers' PUFs deviate from their room temperature values. This would entail, a priori, that a known challenge would not elicit the same response from the device at room temperature and at temperatures exceeding 55 °C. As we shall show below, we have devised a strategy to avoid this issue in order to make keys that are resilient to temperature-induced changes in the underlying PUFs.

5.2.2 *Fingerprint and Key Generation*

Authentication keys are generated by first obtaining a set of digital fingerprints from the imager's PUF, i.e., from the DCM. A flowchart of the procedure used is shown in Fig. 5.4. First, the DCM is normalized. From the normalized DCM, using a Wiener filter and a direct cosine transform (DCT), a high-frequency image is extracted [20]. This component is equivalent to the dark signal nonuniformity [35] map of the imager. In other words, it represents the variation in DCR from pixel to pixel, not

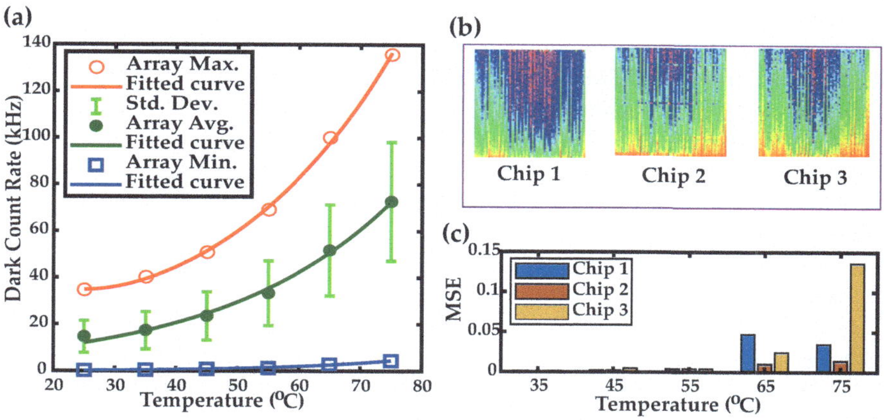

Fig. 5.3 Normalized dark count maps (DCMs) and the effect of temperature. (**a**) Pixel dark counts increase exponentially with temperature, as evidenced by the change in dark count rate as a function of temperature. (**b**) Each chip has a distinct fixed-pattern noise distribution, which results in unique DCMs. (**c**) The mean square error (MSE) between DCMs at different temperatures suggests that there are relatively few changes in the normalized DCM up to 55 °C. Above that temperature threshold, the DCMs change significantly; this suggests that the chips' PUFs are altered at elevated temperature

Fig. 5.4 Flowchart for generating binary fingerprints from the normalized dark count map (DCM), which is the imager's PUF. Two fingerprints are generated (F1 and F2), and they are both used to construct authentication keys

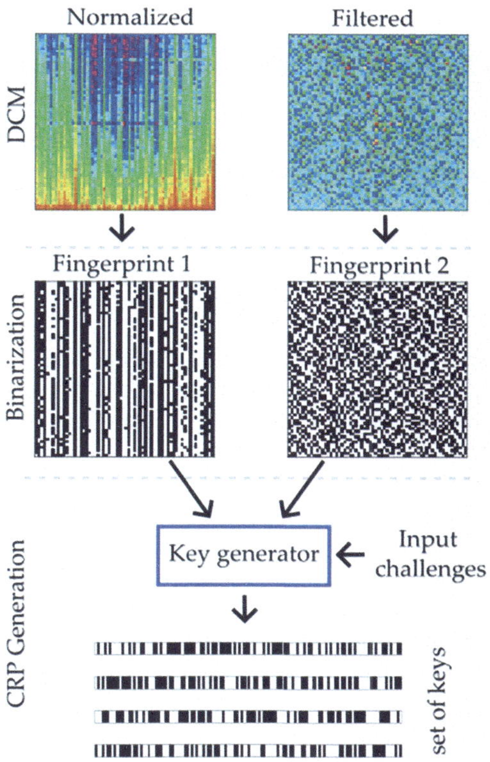

including readout fixed-pattern noise components, which have relatively low spatial frequency. Binary fingerprints, fingerprint 1 (F1) and fingerprint 2 (F2), respectively, are generated from the DCMs using their average value as a threshold.

In contrast with F2, the fingerprint F1 includes low-frequency fixed-pattern noise components associated with the readout chain; these latter components are less dependent on temperature as they are associated with the readout architecture. On the other hand, F2 being related solely to the SPAD's DSNU is strongly dependent on temperature. Thus, we posit that a key that combines both F2 and F1 should be resilient to temperature (by virtue of F1) and unique from chip to chip (by virtue of F2) at the same time.

In the CRP generation step, keys from each fingerprints are generated according to the procedure described below, and they are combined to create a higher-level key that is based on both fingerprints. To generate the keys, each row from F1 and F2 is divided into nonoverlapping blocks with block length L_b. Each pixel p_{ij} in a block j is compared against the block average and assigned a binary value as follows. The keys are then mixed, e.g., simply by appending the two keys generated from the two fingerprints:

$$\begin{aligned} p_{ij} &= 1;\ if\ p_{ij} > mean(p_j) \\ &= 0;\ else \end{aligned} \tag{5.1}$$

5.3 Experimental Validation and Discussion

Various sets of keys were generated, and the uniqueness between key pairs was assessed by calculating the normalized Hamming distance (nHD) between the two keys (k_x and k_y) in a pair as shown in Eq. 5.2. nHD was also used to characterize the robustness of a key against changes in temperature:

$$nHD_{k_x,k_y} = sum(xor(k_x, k_y))/L_k \leq 1 \tag{5.2}$$

Here, $xor(.)$ is the exclusive-OR function that returns the logical vector denoting the instances of different bit pairs. All bit mismatches are summed by the $sum(.)$ function and normalized by the key length L_k. If a generated key is stable, its nHD from itself at another temperature is expected to be 0. In all other cases, e.g., different chip-same challenge or same chip-different challenge, we expect the nHD to be away from the zero line, ideally around 0.5 [9], with sufficient margin to conclude that a key pair includes different keys.

All keys were generated by operating the imagers with 0 V on the perimeter gate, and keys of different lengths were studied; as shown in Fig. 5.5 for keys generated from the same chip with different challenges, longer bit-length keys were more likely to achieve a tighter spread around a nHD of 0.5, which suggests that

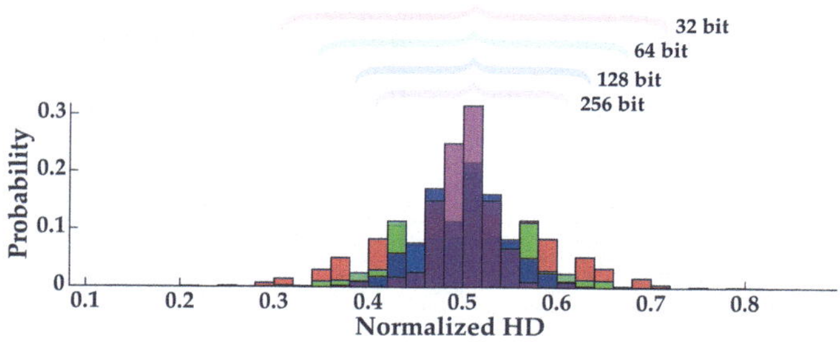

Fig. 5.5 Distribution of normalized Hamming distances for key pairs from the same chip as a function of key bit lengths (32, 64, 128, and 256). Longer keys achieve a tighter spread around 0.5, suggesting increased robustness in differentiation

longer keys are more likely to achieve a sufficiently large margin from the nHD = 0 line [20].

Figure 5.6a shows that a population of short bit-length keys (32 bits) exhibits a large variance around the average nHD (marked by the solid lines through the respective clusters), exacerbated by the increased temperature. In contrast, longer bit-length keys (256 bits), displayed in Fig. 5.6b, show much lower variance around the two cluster means, even at elevated temperatures. This suggests that the long keys are more stable with temperature increases and that key pairs from this population can maintain sufficient difference when temperature is increased. This also points out the benefit of mixing keys from two fingerprints and increasing key length. Such an approach provides a means for small-scale imagers to generate unique and robust keys without being limited to the small number of intrinsic permutations.

Figure 5.6c shows the normalized HD distribution for the 128-bit keys generated using the same challenge but from three different chips. These data show that keys generated from chip to chip are different despite the challenge being the same. This is possible because each sensor is a distinct PUF. While elevated temperature may broaden this distribution, we have found that the resulting distributions are still far from the average nHD (i.e., 0.1) of the same board-same challenge cluster. This suggests that temperature increases do not impede the ability to distinguish one sensor from the other.

Furthermore, we show in Fig. 5.7 that perimeter gating can alter the underlying fingerprints. That is, dark counts across the array may be gradually reduced and eventually brought to vanishing levels with high perimeter gate voltages [36]. As such, the two fingerprints (F1 and F2) that are used to construct keys can be markedly altered or they can be hidden altogether by bringing the perimeter gate voltage magnitude to 7 V. In these high perimeter gating regimes, the device may operate as a high-SNR imager without risking the device's DSNU, and consequently fingerprint F2, to be extracted from images captured with the camera. This is in stark contrast with previous image sensor PUFs in which the underlying DSNU may be characterized from images acquired with the camera [33].

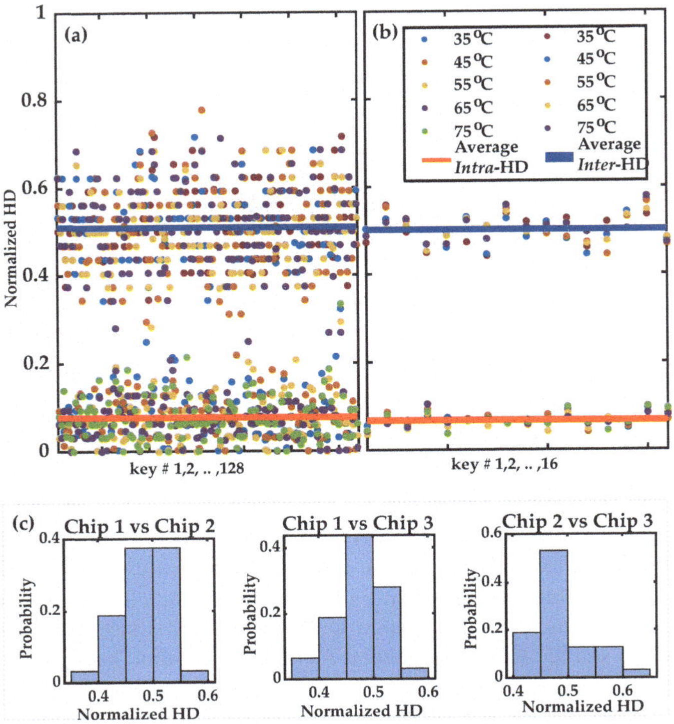

Fig. 5.6 A study of key resilience to temperature changes for key populations generated from the same chip. The *intra*-key normalized Hamming distance (nHD) is the normalized HD between the key and itself. The *inter*-key HD is the normalized HD between two different keys. (**a**) Temperature data for a population of 128 short bit-length keys (32 bits) and (**b**) for a population of 16 longer bit-length keys (256 bits). Distribution of normalized HDs for 128-bit keys generated from the same challenge from chip to chip. The data show that the generated keys are distinct (centered around a normalized HD of 0.5) (**c**)

The ability of perimeter gating to reduce the dark counts and obfuscate a chip's fingerprints generated from its PUF offers a unique opportunity. As shown in Fig. 5.7, the imager can operate in the higher perimeter gate regime during imaging operation when the PUF is concealed. Hence, an adversary will not be able to extract the PUF information from the shared images. On the other hand, the imager can operate in the low-perimeter gate voltage regime to reconfigure its PUF in order to create alternate fingerprints.

5.4 Challenge Response Pair Enhancement

Referring to Fig. 5.1, in an enrollment phase, the key is saved in a remote trusted device like the KDC, and subsequent issuance of the secret challenge is used for authentication. The number of possible unique challenges and corresponding unique

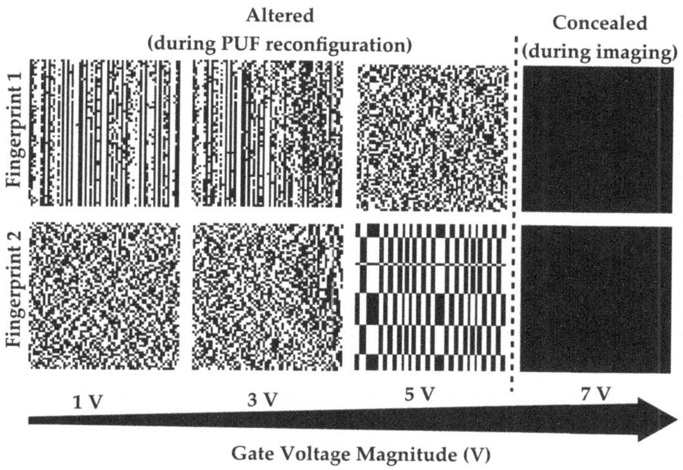

Fig. 5.7 Effect of perimeter gate voltage magnitude on fingerprints F1 and F2

responses forms a challenge-response pair (CRP) space, and the size of this space determines and conditions the strength of the underlying PUF [37].

In this part of the chapter, we demonstrate a novel technique for increasing the CRP space of the aforementioned pg-SPAD-based PUF. Specifically, we show that perimeter gating may be used to dynamically alter the PUF and thus increase the number of unique challenge-response pairs. We show that this technique can overcome the typical size limitations of the CRP space of conventional CMOS imager-based PUFs [33].

5.4.1 Limitations of State-of-the-Art Imager PUFs

In conventional CMOS imager-based PUFs, a challenge vector may be a set of pixel addresses and a response may be a vector of measurement values at each pixel in the challenge vector. For example, the measurements may be the reset voltages, V_{rst}, at each pixel. In this construct, although the number of possible challenges scales with pixel array size, the actual number of CRPs is limited for several reasons.

For instance, consider a scheme where responses from two adjacent pixels are compared to generate a key bit. Once the two pixels are compared, they cannot be used in another challenge since the outcome of the comparison will always yield the same key bit value. Hence, in such a scheme, the CRP space is limited to $\frac{n}{2}$, where n is an even number representing the total number of available pixels in the array.

In yet another example, when two pixels are chosen at random and their responses are compared to produce a key bit, the number of effective challenges becomes much less than the theoretical maximum, i.e., much less than $C_2^n = \frac{n(n-1)}{2}$.

For instance, among three pixels a, b, and c, if it is known that $V_{rst,a} > V_{rst,b}$ and $V_{rst,b} > V_{rst,c}$, then comparing pixel a and pixel c to produce a key bit is ineffective because the outcome is predictable, i.e., it can be inferred that $V_{rst,a} > V_{rst,c}$.

A more robust key generation approach can be used to circumvent the abovenoted issues. For example, comparing a block of pixels with a reference value, e.g., the block's average reset voltage value, can increase the number of distinct CRPs since this strategy does not explicitly give away pixel-to-pixel response information. Thus, a priori, the block approach significantly increases the CRP space, from C_2^n to $C_l^n = \frac{n!}{(n-l)!l!}$, where $2 < l < n - l$ and l is the block length. However, the block method cannot utilize all the possible challenges available from the permutations of the same set of l pixels. This is because permutations of the same block yield the same responses. This is illustrated in Fig. 5.8.

Here, we consider a 16-pixel reference challenge (Fig. 5.8 (top)) and three distinct permutations of the challenge. The permutated challenges yield seemingly different responses and thus seemingly different keys (middle). However, when the permutated challenges' corresponding responses or keys are remapped according to the order of the pixels in the reference challenge, the resulting remapped keys and responses are identical (Fig. 5.8 (bottom)), barring one or two bit flips due to noise. In fact, the remapped keys and responses are equivalent to that of the reference challenge.

Generally, sorting the pixel addresses of the permutations in ascending or descending order reveals a deterministic response pattern that is the same for each permutation. Therefore, the lack of uniqueness in the responses obtained from permutated challenges significantly reduces the CRP space because a block's permutations cannot be used as a challenge. As shown below, our proposed method

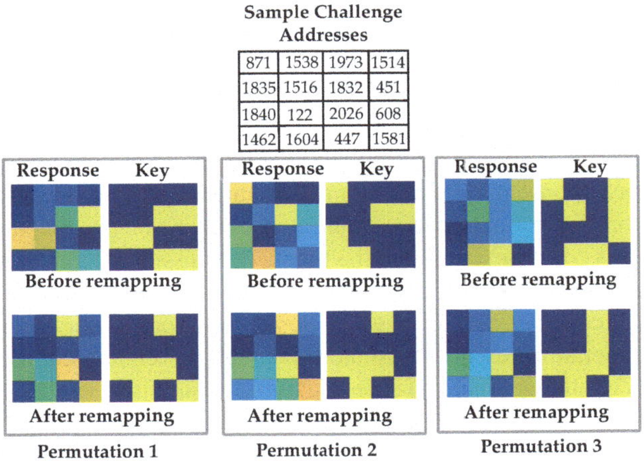

Fig. 5.8 Limitations of conventional CRP generation methods when reusing the same pixel block. (top) Reference challenge. (middle) Normalized dark count response from random permutations of the reference challenge. (bottom) Remapping permutated responses and keys to the reference their reference counterparts

fixes this issue by forcing permutations of the same challenge to yield unique responses.

5.4.2 *Proposed Technique*

We propose a simple technique to make the responses of challenge permutations unique. We do so by assigning pixel position-dependent perimeter gate voltages. For instance, in one exemplary scheme, in a given challenge vector, pixels in odd positions of the vector receive a gate voltage of 0 V while the remaining pixels are set at perimeter gate voltage of 1 V. This effectively changes the responses of the permutations, since different DCMs are invoked when the perimeter gate voltage is altered. As a result, out of $L!$ permutated challenges, $L! - (\frac{L}{2}!)^2 \approx L!$ (assuming $L > 6$) sequences result in unique responses, even after remapping.

Zuo et al. [38] used a similar approach by leveraging the different photoresponses of color pixels in an array, essentially querying three PUFs generated, respectively, from red, green, and blue pixels. In our approach, however, the CRP space can be increased to an even greater number because perimeter gating is analog, which means it can reconfigure the PUF to many more states.

5.5 Measured Results

Here, we present results for an 8-pixel block and a 2-level perimeter gate voltage PUF reconfiguration. The approach easily extends to longer blocks and/or additional PUF reconfiguration levels. Furthermore, one can query the chip with multiple small challenge blocks to mimic a longer one. Similarly, gate voltage assignment can be randomized and increased up to the level at which pixels still exhibit nonzero response.

5.5.1 *CRP Without Perimeter Gate Control*

Figure 5.9a shows the apparently random responses corresponding to the 4 permutations of an 8-pixel reference challenge block. The response at each pixel position ranges from 0 to 60 dark counts. However, after remapping, the responses reveal a deterministic pattern common to all the permutations. This is also evident from the small variance observed for each address, i.e., for each pixel position in the challenge vector (Fig. 5.9b). Hence, without PUF reconfiguration, the CRPs from the permutations of the same set of pixels are a *many-to-one* mapping.

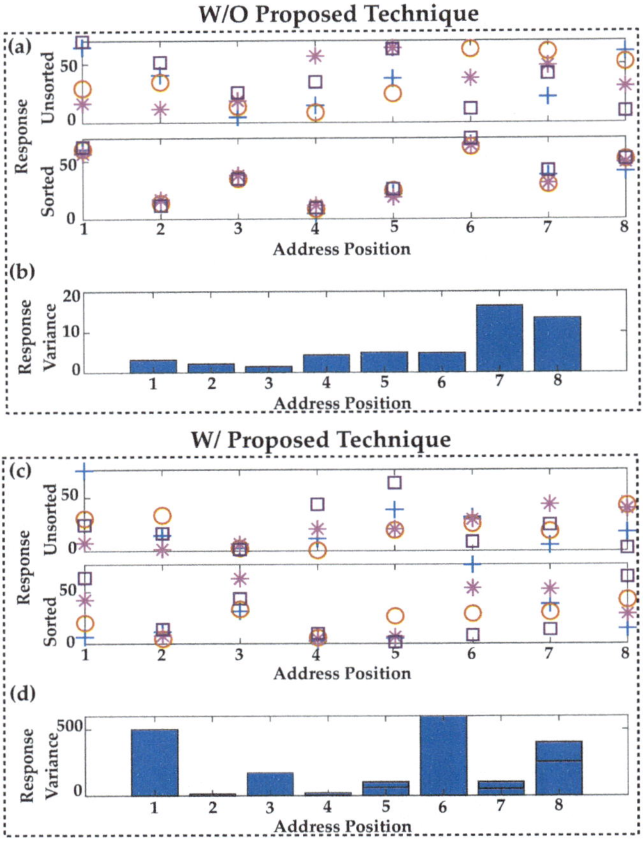

Fig. 5.9 Many-to-one CRP mapping without PUF reconfiguration. (**a**) Responses of permutated challenges before and after remapping. (**b**) Variance of responses for each pixel being in different sequences. (**c**) Unique CRP generation utilizing PUF reconfiguration. Responses of permutated challenges before and after remapping. (**d**) Variance of responses for each pixel being in different sequences

5.5.2 CRP Enhancement with PUF Reconfiguration

Figure 5.9c shows the random responses corresponding to the same four permutations but this time altering perimeter gate voltage based on the position of a pixel in the challenge vector. When remapped, the pixel-wise responses exhibit a higher variance, on average, suggesting that permutation challenges do not yield the same responses.

We note that Fig. 5.9d shows that for some pixels, the variance in the response is small, despite perimeter gate control. We posit that these pixels belong to the least active bins in the dark count histograms and that they are not as sensitive to perimeter gating. Nevertheless, taken as whole, the permutations' responses were different enough to generate distinguishable keys. The proposed method can generate 8-pixel CRPs in the order of 10^{27} from the 4,096 pixels. In contrast, in adjacent pixel pair, random pixel pair, or block method approaches, the CRP space is in the order of 10^3, 10^6, and 10^{23}, respectively.

It should be noted that the total number of CRPs in the proposed method scales exponentially with the increase in the total number of pixels and the block length. However, with the increase of the block length, the computational overhead increases as well. Hence, we suggest to have multiples of smaller blocks as a compromise between complexity and number of available CRPs.

5.5.3 Key Generation and Quality Assessment

We generated keys and evaluated them using nHD as described in the preceding part. Figure 5.10a shows the nHD distribution for all possible 8-bit keys with respect to a randomly selected reference key. As expected, the average nHD is 0.5, with some keys being very close to the reference key. Figure 5.10b shows the histogram for the keys generated from the proposed method from 128 permutated challenges. The resulting distribution of nHDs shows a shift towards an nHD of 1, suggesting that most of the keys are distinct from the reference key, which confirms that permutations of the same block need not be discarded at the same rate as that of the state of the art.

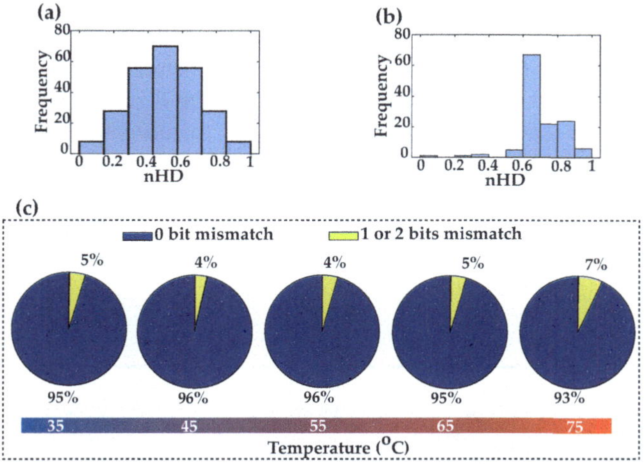

Fig. 5.10 Histogram of the normalized Hamming distances between keys. (a) Histogram for all possible 8-bit keys generated without PUF reconfiguration. (b) Histogram for all possible 8-bit keys generated with PUF reconfiguration

5.5.4 *Temperature Resiliency*

Figure 5.10c displays the temperature resiliency of the generated keys. Since dark count response increases with the increasing temperature, it is expected that the block response will also change at the elevated temperature. However, comparison against the block average counters this since the average response also increases as well. The charts for each of the three chips show that the nHD between the keys generated from the same explicit-implicit challenge results in a stable key over the tested temperature range (27 °C to 75 °C). We found a maximum number of bit flips of 2, which happened 0.8% of the time with average nHD at 0.006 for the same key at different temperatures.

5.6 True Random Number Generation Using pg-SPADs

Random number generators (RNGs) are used in various technologically relevant applications [39–41]. An RNG consists of an algorithm, a device, or a combination thereof that produces sequences of random numbers [42]. A binary RNG produces random bits which are expected to contain no discernible patterns or biases [43]. For instance, the sequence *01100110* shows a repetitive pattern (*0110*), while the sequence *01001000* is heavily biased because it includes significantly more zeroes than ones. Therefore, three components are needed to create a *true* random number generator (TRNG) (see Fig. 5.11). They are a physical entropy source to ensure unpredictability, a transducer to generate electrical signals based on the entropy source, and a post-processor for debiasing the raw bit sequences [44–46].

TRNGs are not susceptible to modeling attacks unlike pseudo-RNGs (PRNGs) which use seeds and deterministic algorithms to generate pseudo-random numbers [47]. Nevertheless, one practical limitation of TRNGs is their lower generation rates compared to that PRNGs [48]. As a result, to mitigate the limitations of both, a new practice has emerged: a high-speed PRNG is used to yield high bit generation rates while a TRNG is used to source random seeds for the PRNG dynamically [49]. Hence, the overall performance of such a system depends on the TRNG's quality; the higher its quality, the more one can relax the constraints on the PRNG's debiasing computational complexity.

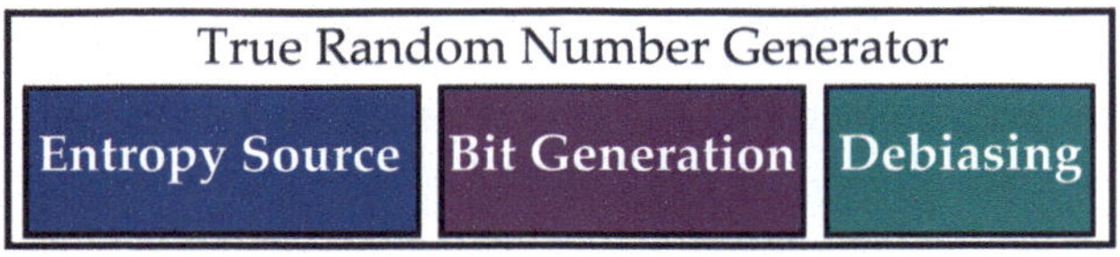

Fig. 5.11 The primary constituent blocks of a true random number generator (TRNG) include a source for randomness, a transducer to generate raw bits, and a debiasing protocol to extract random bits

TRNGs may be implemented using optical quantum RNGs that use photodetectors to generate random bits from transduced photons [50]. They typically use a single-photon avalanche diode (SPAD) as their transducer module. The SPAD is used with a photon source whose light is attenuated and coupled to the detector. It produces a bit stream that is temporally correlated with photon arrival times. The randomness is imparted to the stream by the inherent stochastic nature of the photon arrival process [51–53].

In yet another TRNG implementation that uses SPADs, the detector's noise in the dark may be used as an entropy source [54, 55]. This is in line with the approach featured herein where we use a perimeter-gated SPAD (pg-SPAD) in the dark to generate random binary sequences. Unlike a typical SPAD, a pg-SPAD is equipped with a gate at its perimeter; when this gate is energized with a voltage, the pg-SPAD's dark count rate is altered [25, 56–58]. In sum, in a pg-SPAD, the perimeter gate voltage modulates the rate of events that are observed in the dark.

Conventional optical quantum RNGs have several drawbacks. For instance, they have significant hardware overhead because they must integrate a photon source, optical filters for light attenuation, and in addition, they must include a feedback system to maintain a constant optical power during operation [59]. For photodetector arrays, additional constraints on uniform illumination or photoresponse are imposed to minimize bit bias variations with potentially fluctuating optical power [60]. Similarly, typical dark noise-based SPAD RNGs require hardware for maintaining a constant temperature in order to control bit bias variations with respect to temperature changes [34, 61].

In this part of the chapter, we present pg-SPADs as an alternative to both of the aforementioned approaches. Specifically, our pg-SPAD TRNG does not utilize complex hardware or complex algorithms for bit bias control, nor does it require complex feedback circuits. We benchmarked our pg-SPAD-based TRNG against the statistical test suite (STS) from the National Institute of Standards and Technology (NIST) [62], and it passed all of the 16 required randomness tests from the suite. Further, we show that the pg-SPAD TRNG is robust across temperatures ranging from 27 °C to 75 °C. Specifically, we show that perimeter gating obviates the need for temperature stabilization hardware as it allows the dark count rate to remain the same across the tested temperature range.

5.7 Related Works

In this section, we present a brief critical review of three representative works featuring the implementation of TRNGs using SPADs and their corresponding bit debiasing algorithms. First, we discuss the work of Burri et al. which featured a photon-pumped SPAD-based TRNG. Second, we discuss the computational complexity of the debiasing algorithm used by Stanco et al. Lastly, we discuss the work of Tawfeeq who implemented a dark event-based SPAD TRNG.

Burri et al. demonstrated a photon-based TRNG using a SPAD array [51]. They tuned the SPADs' excess bias voltage and the pulse width of an excitation laser independently until a 50% photon detection probability was reached. Subsequently, they read out the binary frames from the SPAD array and debiased them using Von Neumann's algorithm [63]. This resulted in a random bit efficiency (RBE) close to the theoretical limit of 25%.

Furthermore, although the authors demonstrated that their TRNG was temperature-insensitive, this observation could have stemmed from the fact that the impinging photon rate overwhelmed the temperature-sensitive dark counts. This means at small photon rates, temperature insensitivity would likely not be preserved. More fundamentally, the work of Burri et al. shows that photon count-based designs trade-off device footprint for high throughput since they need to couple the SPAD with a photon source.

To further increase debiasing efficiency, Stanco et al. implemented Peres' algorithm with their photon-based SPAD TRNG [64, 65]. The authors acknowledged the nontrivial computation cost incurred by both the system's processor and memory to implement this algorithm in real time. Further computational costs were incurred with preprocessing steps needed to address non-idealities caused by the device's dead time and its afterpulsing probability. Despite this complexity, the authors achieved an entropy of 1.00 with an optimal clock sampling rate.

Unlike the TRNGs mentioned above, Tawfeeq [55] showed that the dark counts of a SPAD may also be used to source randomness for a TRNG when the SPAD is coupled to a monostable multivibrator. The author showed that by tuning the multivibrator's pulse width, almost equal percentages of zeroes (50.008%) and ones (49.992%) could be achieved. However, one major drawback was the use of a temperature-controlled environment needed to keep the device at 1.5 °C.

Taken together, these works illustrate the shortcomings of the state of the art which our approach directly addresses. Specifically, our approach provides a temperature-stable TRNG that has an RBE within the limit of 25% and an entropy of 1.00 without undue computational and hardware overhead.

5.8 Device Operation

The prototype chip used to implement the novel TRNG included a 64×64 array of actively quenched and actively reset pg-SPADs that was introduced in Chap. 4. The array's dark count rate characteristics and the ability to modulate single-pixel dark count rates with perimeter gating have been described extensively in Ref. [36].

Here, we briefly describe the prototype's pixel architecture and its all-digital readout scheme in the context of a TRNG. Figure 5.12a and b shows a simplified view of the readout architecture employed and a timing diagram of operation, respectively. By asserting SEL at t_0, we start precharging the pg-SPAD's cathode to V_R for a duration of t_{pre}. At the same time, the counter (C) is reset. Beginning at t_1, the device enters the active window and is ready to detect a dark event. V_A maintains

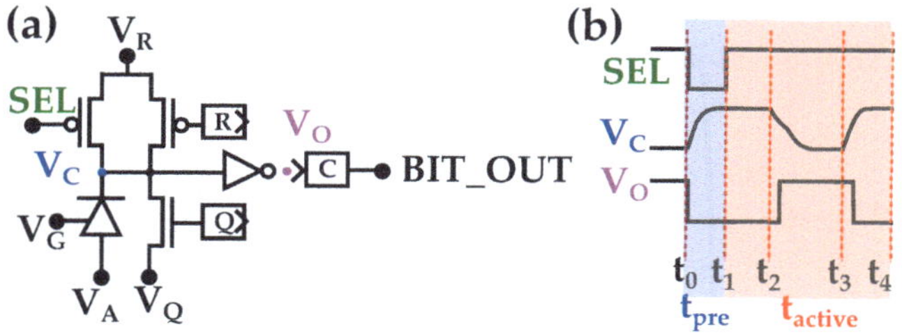

Fig. 5.12 (**a**) Simplified block diagram and (**b**) a timing diagram for operation

the necessary excess bias, V_{ex}, while V_G controls the average dark carrier generation rate [36]. At t_2, the device detects an avalanche and starts to quench passively until V_O toggles and initiates active quenching through the quench signal generator (Q). At the same time, the counter (C) registers the event. After a predetermined delay, the reset signal generator (R) starts resetting the device at t_3. At the end of the active window, the counter is reset at t_4 following a readout by an external data acquisition system. If no dark event occurred within t_{active}, the counter will read zero.

This cycle is repeated continuously and a 0 is produced if no event takes place. Otherwise, a 1 is produced. The raw-bit generation rate is the inverse of ($t_{pre} + t_{active}$), which can be controlled independently.

5.9 Implementation and Measured Results

5.9.1 *Perimeter Gating Achieves Equiprobable Binary States*

Similarly to the number of detected photons within a fixed duration, the number of dark events also follows a Poisson distribution [61]. Hence, the probability, p, of observing no dark events within $t_{active} = t$ is given by

$$p(\lambda(V_G), t) = e^{-\lambda t} \tag{5.3}$$

where λ is the average dark count rate that can be actively controlled by applying a negative voltage of magnitude V_G on the perimeter gate.

As shown in Fig. 5.13a, it is possible to tune the gate voltage, depending on the active window duration t, to achieve $p(\lambda(V_G), t) \rightarrow 0.5$. Under this condition, generating a 1 or a 0 is equally probable. Intuitively, this condition also ensures maximum entropy as it ensures that the pg-SPAD TRNG is not biased towards one of its two possible binary states.

The latter statement is corroborated quantitatively in Fig. 5.13b. Known as the binary entropy function, $H(X)$ shows the entropy of a Bernoulli random variable X against the probability of obtaining a 0, i.e., $P(X = 0)$. A similar plot can be

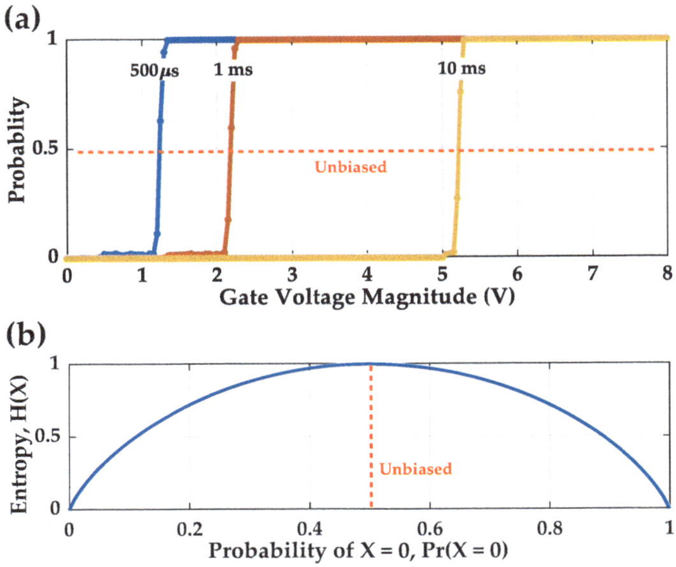

Fig. 5.13 Modulation of the probability of observing no dark event as a function of perimeter gate voltage magnitude and at different activation times (**a**). Entropy of a Bernoulli trial as a function of binary outcome probability. The maximum entropy is observed at the equiprobable (unbiased) condition (**b**)

obtained for $P(X = 1)$. The maximum entropy is achieved when the probability of either one of the binary states is 0.5.

For a random number generator, maximum entropy is desired to ensure the lowest possible bias in the resulting bit stream; this maximizes the TRNG's RBE which is defined as the ratio of the number of extracted bits in the debiased sequence to the number bits routed to the debiasing processor [51]. As such, we used the RBE metric to evaluate the performance of perimeter gating in generating quality raw bit sequences.

5.9.2 *Experimental Methods*

We selected 500 µs as the active window and 10 µs as the precharge time. This yields a nominal raw bit rate of $\simeq$ 2 kHz. V_{ex} was set to 2.5 V, and we deduced the optimal perimeter gate voltage magnitude to be $V_G = 1.24$ V based on the plots shown in Fig. 5.13.

In practice, the raw bit sequences can still exhibit some bias even when the perimeter gate voltage is set at the equiprobable point. This may be due to the

Table 5.1 Von Neumann's algorithm for debiasing binary sequences

In raw data (bit pair)	Replace/discard	In debiased sequence (bit)
00	discard	–
01	replace	0
10	replace	1
11	discard	–

intrinsic non-idealities of the pg-SPAD devices such as their nonzero afterpulsing probabilities. Hence, we applied Von Neumann's algorithm to debias the resulting bit stream [63]. The goal of this algorithm is to test nonoverlapping bit pairs and discard/replace them as described in Table 5.1. Since half of the possible bit pairs are discarded while the remaining bit pairs become a single digit, the theoretical RBE for this classical algorithm is limited to 25% [51].

The raw bits were recorded at 27 °C using a data acquisition system and processed using custom routines written in Matlab. To show the effect of perimeter gating at elevated temperatures, the dark count rate (DCR) was observed at temperatures ranging from 35 °C to 75 °C. Lastly, the bits' randomness was evaluated using the RBE metric and with an STS application [62].

5.9.3 Bit Distribution and RBE Around the Optimal Gate Voltage

Figure 5.14a displays a stacked bar plot showing the bit percentage distribution for ones and zeroes around the optimal gate voltage magnitude, i.e., around 1.24 V. We see that the percentage of ones gradually reduces to zero with increasing V_G. Conversely, the percentage of zeros gradually vanishes with decreasing V_G. At the optimal point, an equal distribution of ones and zeros is observed, which corresponds to an entropy of 1.00.

Figure 5.14b shows the RBE around the 50% bit distribution point. As expected, the RBE is very low, if not zero, for heavily biased sequences. For instance, for large positive deviations from the optimal voltage, the raw bits become all zeroes due to the device's lack of activity. Conversely, for large negative deviations, the dark count rate is high enough to produce at least one dark event within the activation period, resulting in sequences with all ones. We observed an RBE of 24.8% at the optimal condition, which is very close to the theoretical limit, implying that the raw bits from the pg-SPAD TRNG contain minimal bias. Consequently, we generated 1 megabit of data, and using Von Neumman's algorithm, we produced a debiased bit stream with the perimeter gate set at the optimal voltage of 1.24 V. The degree of randomness in these bits were then evaluated using NIST's STS program, as described below.

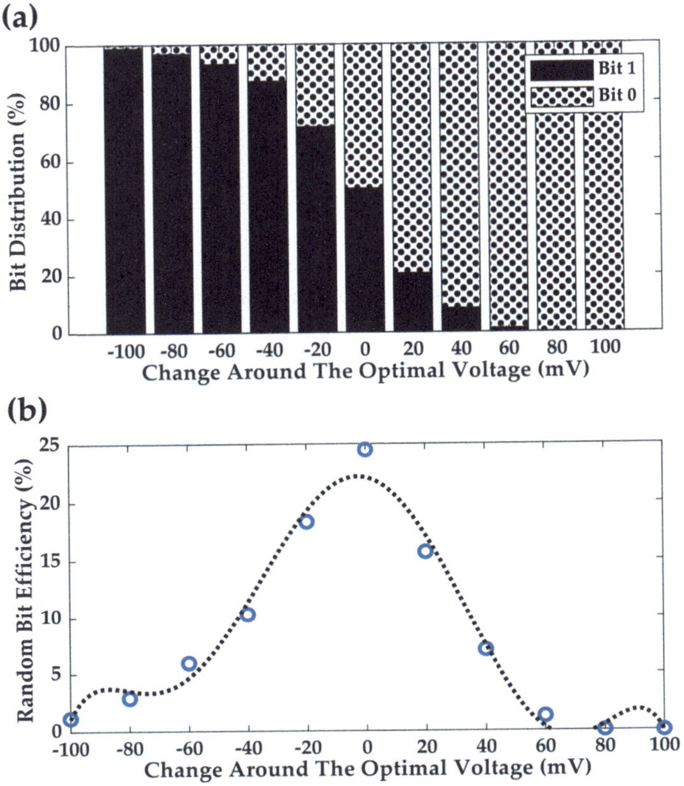

Fig. 5.14 (**a**) Stacked bar plot showing the bit distribution in the raw bit streams at different gate voltages around the optimal operating point. For $V_{ex} = 2.5$ V with $t_{active} = 500\,\mu$s, we measured the optimal gate voltage to be 1.24 V. (**b**) Distribution of random bit efficiency (RBE) with respect to the perimeter gate voltage around the optimal operating point. We achieved a peak RBE of 24.8% at the optimal gate voltage of 1.24 V

5.9.4 NIST Randomness Tests

We administered each of the 16 NIST STS tests to the debiased bit stream. Table 5.2 shows the tests that were conducted along with the p-values obtained from the tests. Here, randomness is confirmed when a p-value is greater than the confidence level of a test, which we set at the default value of $\alpha = 0.01$. Our results revealed that for each of the 16 tests, randomness was confirmed. Furthermore, we note that for the random excursions and random excursion variant tests, we only reported only one of the possible 8 and 18 p-values, respectively. However, the p-values for all the possible states for these tests were also greater than $\alpha = 0.01$.

Table 5.2 NIST STS randomness test results ($\alpha = 0.01$)

No.	Test	*p-Value*
1	Frequency (monobit)	0.537
2	Block frequency (m = 128)	0.856
3	Runs	0.290
4	Longest run of ones in a block	0.941
5	Binary matrix rank	0.325
6	Discrete Fourier transform (spectral)	0.278
7	Nonoverlapping template matching (m = 9)	0.312
8	Overlapping template matching (m = 9)	0.028
9	Maurer's "universal statistical"	0.645
10	Linear complexity (M = 500)	0.348
11	Serial (m = 16, $\nabla\psi_m^2$)	0.416
12	Approximate entropy (m =10)	0.516
13	Cumulative sums (Forward)	0.505
14	Cumulative sums (Backward)	0.618
15	Random excursions (x = +1)	0.717
16	Random excursions Variant (x = −1)	0.298

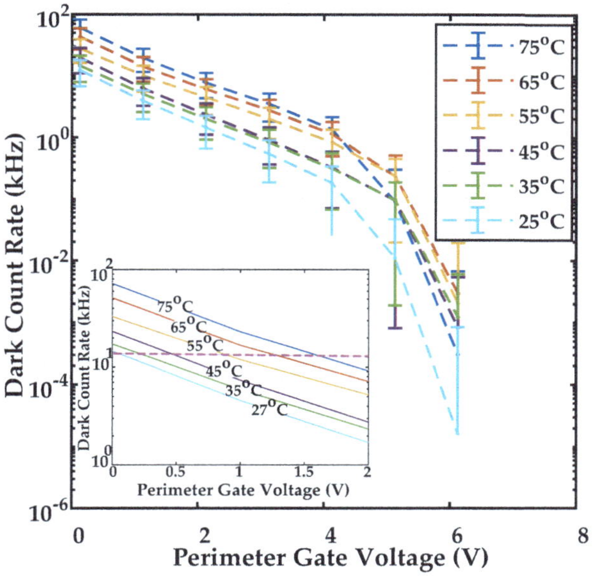

Fig. 5.15 Effect of temperature and perimeter gate voltage on a pg-SPAD's DCR. By adaptively changing V_G, it is possible to maintain a constant DCR across a wide temperature range

5.9.5 Perimeter Gating Compensates for Temperature Change

Lastly, to make a temperature-stable TRNG, we showed that the average dark count rate could be maintained across a wide temperature range. As is well known, a SPAD's DCR increases with temperature [61]. Interestingly, perimeter gating offers a countermeasure for that increase and helps maintain a stable DCR even when temperature is increased.

As shown in Fig. 5.15, our chip's array average DCR decreases exponentially with increased perimeter gate voltage magnitude for temperatures ranging from 27 °C to 75 °C. Therefore, with perimeter gating, it is possible to operate the device with the same average DCR at different temperatures; this is shown by the intersections of the dashed line with the DCR curves (in the inset). These intersection points represent the equiprobable points for the perimeter gate voltage, i.e., the perimeter gate voltage at which entropy is maximized.

5.10 Conclusions

In this chapter, we demonstrated the utility of perimeter-gated single-photon avalanche diode (pg-SPAD) array as building blocks in the hardware security primitives. We have demonstrated a PUF application, CRP enhancement technique, and a true random number generator. We used the devices' dark count rates as the source of spatial and temporal entropy instead of a photon source to reduce system-level complexity. We showed that our pg-SPAD RNG was stable across a wide temperature range; this was achieved by changing the perimeter gate voltage magnitude to keep the average dark count rate constant despite increases in temperature. One of the drawbacks of dark count-based RNGs such as the one presented here is their lower generation rate compared to photon-based RNGs since the incoming photon rates for the latter dominate carrier generation rates. Nevertheless, for high bit rate applications, the proposed pg-SPAD TRNG may be used to provide seeds for a high-speed PRNG thereby simultaneously mitigating modeling attacks and ensuring high bit rates.

References

1. M.S. Sajal, M. Dandin, Concealable physically unclonable functions and key generation using a geiger mode imager, in *2023 IEEE International Symposium on Circuits and Systems (ISCAS)* (IEEE, 2023), pp. 1–5. https://doi.org/10.1109/ISCAS46773.2023.10182123. ISBN 978-1-6654-5109-3
2. M.S. Sajal, M. Dandin, Challenge-response pair space enhancement for imager-based physically unclonable functions, in *2023 IEEE 66th International Midwest Symposium on Circuits and Systems (MWSCAS)* (2023), pp. 217–221. https://doi.org/10.1109/MWSCAS57524.2023.10405869

3. M.S. Sajal, M. Dandin, True random number generation using dark noise modulation of a single-photon avalanche diode. IEEE Trans. Circuits Syst. II: Express Briefs **71**(3), 1586–1590 (2024)
4. Y. Gao, S.F. Al-Sarawi, D. Abbott, Physical unclonable functions. Nat. Electron. **3**(2), 81–91 (2020)
5. T. McGrath, I.E. Bagci, Z.M. Wang, U. Roedig, R.J. Young, A PUF taxonomy. Appl. Phys. Rev. **6**(1), 011303 (2019)
6. A. Shamsoshoara, A. Korenda, F. Afghah, S. Zeadally, A survey on physical unclonable function (PUF)-based security solutions for internet of things. Comput. Netw. **183**, 107593 (2020)
7. A. Balan, T. Balan, M. Cirstea, F. Sandu, A PUF-based cryptographic security solution for IoT systems on chip. EURASIP J. Wirel. Commun. Netw. **2020**(1), 231 (2020)
8. T. Idriss, H. Idriss, M. Bayoumi, A PUF-based paradigm for IoT security, in *2016 IEEE 3rd World Forum on Internet of Things (WF-IoT)* (IEEE, 2016), pp. 700–705. https://doi.org/10.1109/WF-IoT.2016.7845456. ISBN 978-1-5090-4130-5
9. R. Helinski, Evaluating physical unclonable functions. [Online]. Available: https://www.osti.gov/biblio/1766751
10. S. Sultana, M.I. Jabiullah, M.L. Rahman, Improved Needham-Schroeder protocol for secured and efficient key distributions, in *2009 12th International Conference on Computers and Information Technology* (2009), pp. 564–569. https://doi.org/10.1109/ICCIT.2009.5407301
11. L. Lin, D. Holcomb, D.K. Krishnappa, P. Shabadi, W. Burleson, Low-power sub-threshold design of secure physical unclonable functions, in *Proceedings of the 16th ACM/IEEE International Symposium on Low Power Electronics and Design – ISLPED'10* (ACM Press, New York, 2010), p. 43. https://doi.org/10.1145/1840845.1840855. ISBN 9781450301466
12. B. Gassend, D. Clarke, M. van Dijk, S. Devadas, Silicon physical random functions, in *Proceedings of the 9th ACM Conference on Computer and Communications Security – CCS'02* (ACM Press, New York, 2002), p. 148. https://doi.org/10.1145/586110.586132. ISBN 1581136129
13. A. Maiti, J. Casarona, L. McHale, P. Schaumont, A large scale characterization of RO-PUF, in *2010 IEEE International Symposium on Hardware-Oriented Security and Trust (HOST)* (IEEE, 2010), pp. 94–99. https://doi.org/10.1109/HST.2010.5513108. ISBN 978-1-4244-7811-8
14. M.T. Rahman, D. Forte, J. Fahrny, M. Tehranipoor, ARO-PUF: an aging-resistant ring oscillator PUF design, in *Design, Automation & Test in Europe Conference & Exhibition (DATE), 2014* (IEEE Conference Publications, New Jersey, 2014), pp. 1–6. https://doi.org/10.7873/DATE.2014.082. ISBN 9783981537024
15. R. Maes, P. Tuyls, I. Verbauwhede, Low-overhead implementation of a soft decision helper data algorithm for SRAM PUFs (2009), pp. 332–347
16. M. Liu, C. Zhou, Q. Tang, K.K. Parhi, C.H. Kim, A data remanence based approach to generate 100% stable keys from an SRAM physical unclonable function, in *2017 IEEE/ACM International Symposium on Low Power Electronics and Design (ISLPED)* (IEEE, 2017), pp. 1–6. https://doi.org/10.1109/ISLPED.2017.8009192. ISBN 978-1-5090-6023-8
17. Y. Cao, L. Zhang, S.S. Zalivaka, C.-H. Chang, S. Chen, CMOS image sensor based physical unclonable function for coherent sensor-level authentication. IEEE Trans. Circuits Syst. I: Regul. Pap. **62**(11), 2629–2640 (2015)
18. S.-C. Kim, S.-C. Lim, J. Shin, J. Choi, Biometrics for electronic eyes: system authentication with embedded CMOS image sensor. IEEE Trans. Consum. Electron. **62**(3), 209–215 (2016)
19. D. Valsesia, G. Coluccia, T. Bianchi, E. Magli, User authentication via PRNU-based physical unclonable functions. IEEE Trans. Inf. Forensics Secur. **12**(8), 1941–1956 (2017)
20. Y. Kim, Y. Lee, CamPUF: physically unclonable function based on CMOS image sensor fixed pattern noise, in *2018 55th ACM/ESDA/IEEE Design Automation Conference (DAC)* (IEEE, 2018), pp. 1–6. https://doi.org/10.1109/DAC.2018.8465908
21. F. Zappa, S. Tisa, A. Tosi, S. Cova, Principles and features of single-photon avalanche diode arrays. Sensors Actuators A: Phys. **140**(1), 103–112 (2007)

22. B. Aull, Geiger-mode avalanche photodiode arrays integrated to all-digital CMOS circuits. Sensors **16**(4), 495 (2016)
23. Q. Sun, X. Dun, Y. Peng, W. Heidrich, Depth and transient imaging with compressive SPAD array cameras, in *2018 IEEE/CVF Conference on Computer Vision and Pattern Recognition* (IEEE, 2018), pp. 273–282. https://doi.org/10.1109/CVPR.2018.00036. ISBN 978-1-5386-6420-9
24. F. Villa, E. Conca, F. Madonini, F. Zappa, SPAD 3D LiDAR sensors for automotive, industrial automation and surveillance, in *OSA Imaging and Applied Optics Congress 2021 (3D, COSI, DH, ISA, pcAOP)* (Optica Publishing Group, Washington, DC, 2021), p. CTh4E.1. https://doi.org/10.1364/COSI.2021.CTh4E.1. ISBN 978-1-943580-89-7
25. B. Nouri, M. Dandin, P. Abshire, Characterization of single-photon avalanche diodes in standard CMOS, in *2009 IEEE Sensors* (IEEE, 2009), pp. 1889–1892. https://doi.org/10.1109/ICSENS.2009.5398384. ISBN 978-1-4244-4548-6
26. M. Dandin, N. Nelson, V. Saveliev, H. Ji, P. Abshire, I. Weinberg, Single photon avalanche detectors in standard CMOS, in *2007 IEEE Sensors* (IEEE, 2007), pp. 585–588. https://doi.org/10.1109/ICSENS.2007.4388466. ISBN 978-1-4244-1261-7
27. M. Dandin, A. Akturk, B. Nouri, N. Goldsman, P. Abshire, Characterization of single-photon avalanche diodes in a 0.5 μm standard CMOS process – Part 1: Perimeter breakdown suppression. IEEE Sens. J. **10**(11), 1682–1690 (2010)
28. M. Dandin, M.H.U. Habib, B. Nouri, P. Abshire, N. McFarlane, Characterization of single-photon avalanche diodes in a 0.5 μm standard CMOS process—Part 2: equivalent circuit model and geiger mode readout. IEEE Sens. J. **16**(9), 3075–3083 (2016)
29. B. Nouri, M. Dandin, P. Abshire, Large-area low-noise single-photon avalanche diodes in standard CMOS, in *2012 IEEE Sensors* (IEEE, 2012), pp. 1–5. https://doi.org/10.1109/ICSENS.2012.6411365. ISBN 978-1-4577-1767-3
30. M. Dandin, P. Abshire, High signal-to-noise ratio avalanche photodiodes with perimeter field gate and active readout. IEEE Electron Device Lett. **33**(4), 570–572 (2012)
31. A. Akturk, M. Dandin, N. Goldsman, P. Abshire, Modeling of perimeter-gated silicon avalanche diodes fabricated in a standard single-well CMOS process, in *2009 International Semiconductor Device Research Symposium* (IEEE, 2009), pp. 1–2. https://doi.org/10.1109/ISDRS.2009.5378222. ISBN 978-1-4244-6030-4
32. M.S. Sajal, M. Dandin, Concealable physically unclonable functions and key generation using a geiger mode imager, in *2023 IEEE International Symposium on Circuits and Systems (ISCAS)* (2023), pp. 1–5. https://doi.org/10.1109/ISCAS46773.2023.10182123
33. H. Yamada, S. Okura, M. Shirahata, T. Fujino, Modeling attacks against device authentication using CMOS image sensor PUF. IEICE Electron. Express **18**(7), 20210058 (2021)
34. M. Hofbauer, B. Steindl, H. Zimmermann, Temperature dependence of dark count rate and after pulsing of a single-photon avalanche diode with an integrated active quenching circuit in 0.35μm CMOS. J. Sens. **2018**, 1–7 (2018)
35. M. Kłosowski, Y. Sun, Fixed pattern noise reduction and linearity improvement in time-mode CMOS image sensors. Sensors **20**(20), 5921 (2020)
36. M.S. Sajal, K.-C. Lin, B. Senevirathna, S. Lu, M. Dandin, Perimeter-gated single-photon avalanche diode imager with vanishing room temperature dark count probability, in *2022 29th IEEE International Conference on Electronics, Circuits and Systems (ICECS)* (IEEE, 2022), pp. 1–4. https://doi.org/10.1109/ICECS202256217.2022.9970824. ISBN 978-1-6654-8823-5
37. H. Martin, P. Peris-Lopez, G. Natale, M. Taouil, S. Hamdioui, Enhancing PUF based challenge–response sets by exploiting various background noise configurations. Electronics **8**(2), 145 (2019)
38. H. Zuo, Q. Li, H. Zheng, Y. Yang, X. Zhao, An optically-reconfigurable PUF based on logarithmic photoreceptor of CMOS dynamic vision sensors. IEEE Trans. Electron Devices **69**(9), 5395–5398 (2022)
39. T. Lugrin, Random number generator, in *Trends in Data Protection and Encryption Technologies*, ed. by V. Mulder, A. Mermoud, V. Lenders, B. Tellenbach (Springer Nature, Switzerland, 2023), pp. 31–34. https://doi.org/10.1007/978-3-031-33386-6_7

40. K. Seyhan, S. Akleylek, Classification of random number generator applications in IoT: a comprehensive taxonomy. J. Inf. Secur. Appl. **71**, 103365 (2022)
41. M.M. Jacak, P. Jóźwiak, J. Niemczuk, J.E. Jacak, Quantum generators of random numbers. Sci. Rep. **11**(1), 16108 (2021)
42. A. Marghescu, P. Svasta, Into generating true random numbers – a practical approach using FPGA, in *2015 IEEE 21st International Symposium for Design and Technology in Electronic Packaging (SIITME)* (IEEE, 2015), pp. 319–322. https://doi.org/10.1109/SIITME.2015.7342346
43. M. Herrero-Collantes, J.C. Garcia-Escartin, Quantum random number generators. Rev. Modern Phys. **89**(1), 015004 (2017)
44. K. Balasubramanian, Hash Functions and Their Applications (2018), pp. 66–77
45. P. Lacharme, Post-processing functions for a biased physical random number generator, in *Fast Software Encryption*, ed. by K. Nyberg (Springer, Berlin/Heidelberg, 2008), pp. 334–342. https://doi.org/10.1007/978-3-540-71039-4_21
46. M. Dichtl, Bad and good ways of post-processing biased physical random numbers, in *Fast Software Encryption*, ed. by A. Biryukov (Springer, Berlin/Heidelberg, 2007), pp. 137–152. https://doi.org/10.1007/978-3-540-74619-5_9
47. Y. Dodis, D. Pointcheval, S. Ruhault, D. Vergniaud, D. Wichs, *Security Analysis of Pseudo-Random Number Generators with Input: /Dev/Random is Not Robust* (Association for Computing Machinery, 2013), pp. 647–658. https://doi.org/10.1145/2508859.2516653
48. F. James, L. Moneta, Review of high-quality random number generators. Comput. Softw. Big Sci. **4**(1), 2 (2020)
49. C. Ryan, M. Kshirsagar, G. Vaidya, A. Cunningham, R. Sivaraman, Design of a cryptographically secure pseudo random number generator with grammatical evolution. Sci. Rep. **12**(1), 8602 (2022)
50. O. Guillan-Lorenzo, M. Troncoso-Costas, D. Alvarez-Outarelo, F.J. Diaz-Otero, J.C. Garcia-Escartin, Optical quantum random number generators: a comparative study. Opt. Quantum Electron. **55**(2), 185 (2023)
51. S. Burri, D. Stucki, Y. Maruyama, C. Bruschini, E. Charbon, F. Regazzoni, SPADs for quantum random number generators and beyond, in *2014 19th Asia and South Pacific Design Automation Conference (ASP-DAC)* (IEEE, 2014), pp. 788–794. https://doi.org/10.1109/ASPDAC.2014.6742986
52. A. Incoronato, F. Severini, E. Conca, F. Villa, F. Zappa, Linear SPAD array single- and multiple-photon coincidence-based quantum random number generator, in *OSA Optical Sensors and Sensing Congress 2021* (2021). https://doi.org/10.1364/SENSORS.2021.SW5F.3
53. A. Tontini, L. Gasparini, N. Massari, R. Passerone, SPAD-based quantum random number generator with an Nth-order rank algorithm on FPGA. IEEE Trans. Circuits Syst. II: Express Briefs **66**(12), 2067–2071 (2019)
54. J. Lin, Y. Wang, Q. Cao, J. Kuang, L. Wang, True random number generation based on arrival time and position of dark counts in a multichannel silicon photomultiplier. Rev. Sci. Instrum. **90**(11), 114704 (2019)
55. S. Tawfeeq, A random number generator based on single-photon avalanche photodiode dark counts. J. Lightwave Technol. **27**(24), 5665–5667 (2009)
56. M. Dandin, N. Nelson, V. Saveliev, H. Ji, P. Abshire, I. Weinberg, Single photon avalanche detectors in standard CMOS, in *2007 IEEE Sensors* (IEEE, 2007), pp. 585–588. https://doi.org/10.1109/ICSENS.2007.4388466
57. M. Dandin, A. Akturk, B. Nouri, N. Goldsman, P. Abshire, Characterization of single-photon avalanche diodes in a 0.5 μm standard CMOS process–part 1: perimeter breakdown suppression. IEEE Sens. J. **10**(11), 1682–1690 (2010)
58. M. Dandin, M.H.U. Habib, B. Nouri, P. Abshire, N. McFarlane, Characterization of single-photon avalanche diodes in a 0.5 μm standard CMOS process–part 2: equivalent circuit model and geiger mode readout. IEEE Sens. J. **16**(9), 3075–3083 (2016)

59. G. Verschaffelt, M. Khoder, G. Van der Sande, Random number generator based on an integrated laser with on-chip optical feedback. Chaos: Interdiscip. J. Nonlinear Sci. **27**(11), 114310 (2017)
60. B.K. Park, H. Park, Y.-S. Kim, J.-S. Kang, Y. Yeom, C. Ye, S. Moon, S.-W. Han, Practical true random number generator using CMOS image sensor dark noise. IEEE Access **7**, 91407–91413 (2019)
61. Y. Kang, H.X. Lu, Y.-H. Lo, D.S. Bethune, W.P. Risk, Dark count probability and quantum efficiency of avalanche photodiodes for single-photon detection. Appl. Phys. Lett. **83**(14), 2955–2957 (2003)
62. L. Bassham, A. Rukhin, J. Soto, J. Nechvatal, M. Smid, S. Leigh, M. Levenson, M. Vangel, N. Heckert, D. Banks, A statistical test suite for random and pseudorandom number generators for cryptographic applications, in *NIST Special Publication 800-22* (2010)
63. J. Von Neumann, Various techniques used in connection with random digits, Applied Math Series, Notes by G. E. Forsythe, in *National Bureau of Standards*, vol. 12 (NIST, 1951), pp. 36–38
64. A. Stanco, D.G. Marangon, G. Vallone, S. Burri, E. Charbon, P. Villoresi, Efficient random number generation techniques for CMOS single-photon avalanche diode array exploiting fast time tagging units. Phys. Rev. Res. **2**(2), 023287 (2020)
65. Y. Peres, Iterating Von Neumann's procedure for extracting random bits. Ann. Stat. **20**(1), 590–597 (1992)

Chapter 6
Silicon Photomultipliers

6.1 Introduction

Silicon photomultipliers (SiPM) are less bulky and less expensive, they have lower voltage operation, and they are insensitive to magnetic fields; these attributes make them a viable alternative to photomultiplier tubes (PMTs). An SiPM sums the output current of multiple single-photon avalanche diode (SPAD) microcells, each of which is in series with a quenching resistor (Fig. 6.1) [2, 27–30]. SiPMs, being comprised of SPADs, can be used in many of the same applications in which SPADs are used. Compared to a single SPAD, they use more silicon area, but they offer many other advantages. Applications span the vast uses of imaging and are used for exploration in light detection and ranging (LiDAR) applications, detection of nuclear particles such as neutrons and gamma particles (e.g., in positron emission tomography in the medical field and neutron scattering experiments in high-energy physics), and detecting entangled photons in quantum applications [25, 26]. SiPMs can be classified into either analog SiPMs or digital SiPMs, and the choice of SiPM is highly dependent on the application and quantity, such as energy or timing, desired to be observed.

Being comprised of SPADs, SiPMs will show similar noise processes to SPADs. Thus, in the dark an SiPM will have a dark current, which is the sum of n SPADs avalanching randomly, according to the statistics of the device and the underlying noise process. The primary sources of noise are trap-assisted tunneling, band-to-band tunneling, and carrier generation through thermal events. Use of the perimeter-gated approach, being a field-induced effect, has an impact on the non-photon-generated carriers. This has the overall effect of improving the signal-to-noise ratio of the SiPM in the visible spectral range (350–800 nm) [2, 28].

M. Dandin et al., *Single-Photon Avalanche Diodes and Photon Counting Systems*,
https://doi.org/10.1007/978-3-031-64334-7_6

6.2 18 x 18 Analog SiPM Array

An 18 × 18 SiPM, using perimeter-gated SPADs and a transistor in triode as the quenching resistor, is shown in Fig. 6.1, with the photomicrograph being shown in Fig. 6.2 [14]. All the gates of the pg-SPAD and transistor are connected to a common node. In addition, the anodes are tied to a common bias voltage. Each device can thus undergo an individual avalanche and quench process that is independent of the other devices. The quenching resistor of a SPAD is typically on the order of 100 kΩ. However, for many processes without a high-resistance implant, high-value resistors can occupy a large area leading to a poor fill factor. That is, the active optical area is reduced. This motivates the use of transistors. In Fig. 6.1 a p-type MOSFET is used. The voltage bias across the SPAD in series with the resistor is typically above the breakdown voltage of the SPAD. This is significantly higher than the rated voltage of the transistors. Thus, care must be taken in applying the appropriate gate voltage to the transistor to mitigate any possibility of oxide breakdown and the junction voltages of the transistor must be carefully considered. The total current from multiple avalanching SPADs are summed at the output, and as the luminance increases, the number of avalanching SPADs increases and the total output current

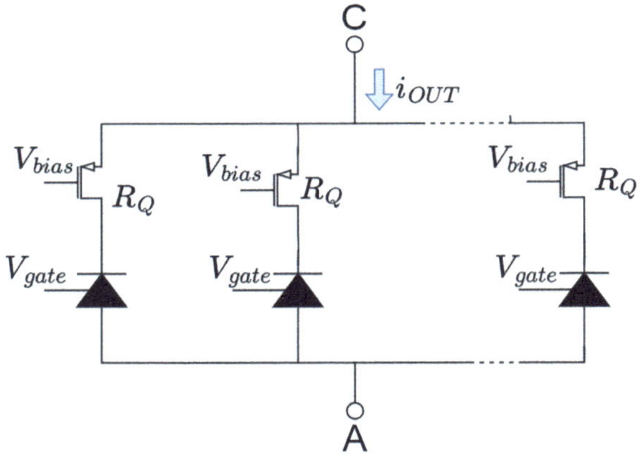

Fig. 6.1 An 18 × 18 analog SiPM array [14]

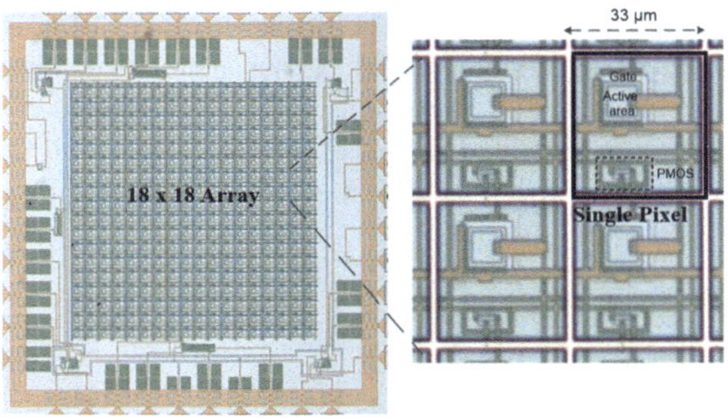

Fig. 6.2 Photomicrograph of an 18 × 18 analog SiPM array [14]

increases. It should be noted that even under uniform illumination, each SPAD has a finite probability of avalanching in a given time window when hit with a photon.

The actual junction in Fig. 6.2 is implemented using the p diffusion and n-well layers in a 0.5-μm process. While corners produce higher fields, leading to shapes such as hexagons and circles being the ideal form factor, the limitations of the design rules in standard CMOS, particularly in newer technologies, may strictly prohibit these options. Thus, due to these limitations, the SPAD, regardless of material, may be square or rectangular shaped. In a typical n-well process, the junctions can be p^+/n-well, n-well-psub, or n^+-psub. Each layer has a different depth and doping levels and thus gives rise to a range of breakdown voltages for the SPAD. The breakdown voltage can range across typical junctions as just described from 10 V up to 22 V. Since polysilicon will block light, the use of the pg-SPAD does reduce the total active area, and there is a trade-off between the width of the polysilicon and the rest of the junction. For Fig. 6.2, in one dimension, the active area is 1/3 of the total area.

6.3 pg-SPAD SiPM Characterization

Pg-SPAD SiPMs have been characterized electrically, optically, and as a function of temperature. Significantly, the additional polysilicon gate allows the performance characteristics, specifically the signal-to-noise ratio and sensitivity, to be modulated as a function of applied gate bias. In Ref. [14], three sample pg-SPAD SiPM chips were tested, and the average value was reported. It was noted that the maximum standard deviation from the mean was on the order of $\sim$9%.

6.3.1 pg-SPAD Breakdown Voltage

Pg-SPAD devices have been implemented in multiple processes including a long-channel CMOS process, a submicron process, and a high-voltage CMOS process. The devices are on the order of $22 \times 22\,\mu$m and use similar junction types. I-V characterization was performed using high-accuracy source measure units, where one unit varies the applied voltage across the device moving it from forward to reverse to reverse breakdown regions of operation, while the other unit varies the applied gate bias. The gate voltage of the transistor is kept constant to provide a roughly 100-kΩ size quenching resistor. A crude estimate of the breakdown voltage, V_{Br}, can be derived at simply examining the I-V characteristic and taking the voltage at which the current exceeds some nominal current by a predefined threshold value. However, a more accurate estimate can be derived by considering the voltage at which maximum curvature occurs in the reverse breakdown region. It should be noted that there can also be a "second breakdown" point in the reverse bias characteristic. For the pg-SPAD, the breakdown voltage as a function of gate

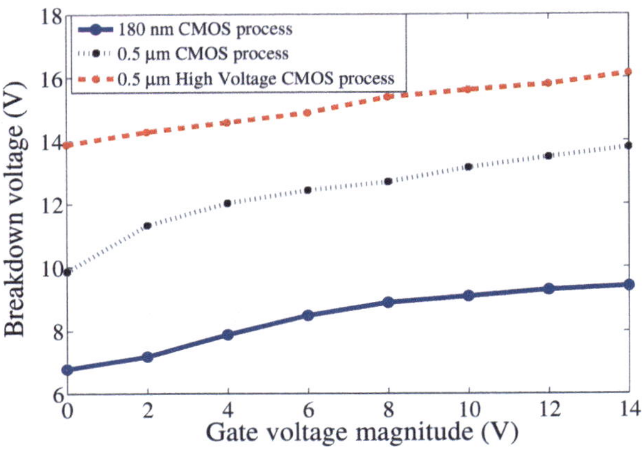

Fig. 6.3 Experimentally measured breakdown voltage, V_{Br}, as a function of the applied gate bias voltage, $|V_G|$. The pg-SPAD device size is $22 \times 22\,\mu$m across three unique technologies [14]

voltage can be seen in Fig. 6.3, where changing the gate voltage results in an increase in the breakdown voltage regardless of process. The exact total change variation is a function of the junction type and doping level. The effect has also been confirmed for newer technologies which incorporate shallow trench isolation.

6.3.2 pg-SPAD SiPM I-V Characterization

The gate voltage, which is an applied field effect on the SPAD, ultimately affects the noise processes of the device. The excess bias voltage also influences the noise of the device and there is a trade-off in choosing the optimum bias voltages for a given application. The excess bias increases the drift velocity of charge carriers which have been trapped due to defects and which have been generated through thermal processes. This has the overall effect of increasing the dark current of the device. The gate voltage, on the other hand, decreases the dark current. The gate voltage creates a more uniform electric field across the junction and at a basic level can be thought of as modulating the availability of charge carriers or the energy band diagram of the structure. The probability of an avalanche, regardless of the source of the carrier, is inversely proportional to the gate bias, resulting in reduced dark current levels as seen in Figs. 6.4 and 6.5.

6.3.3 pg-SPAD SiPM: Optical Characterization

Figure 6.6 shows an optical test setup for characterizing the spectral response of a pg-SPAD SiPM in [14]. Typical characterization requires an optical power meter to

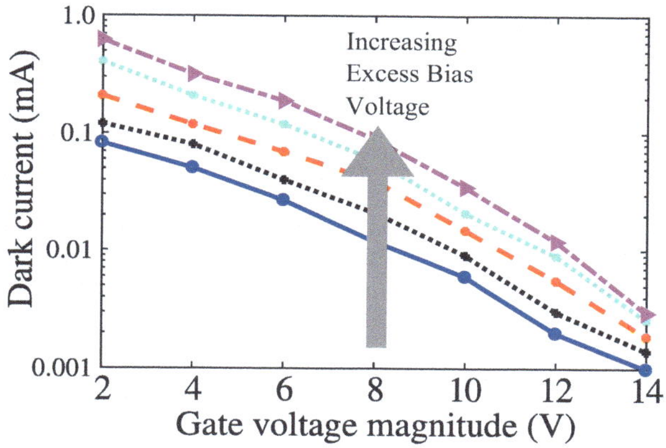

Fig. 6.4 pg-SPAD SiPM dark current as a function of gate voltage, $|V_G|$, for varying excess bias voltage, V_{Exc} [14]

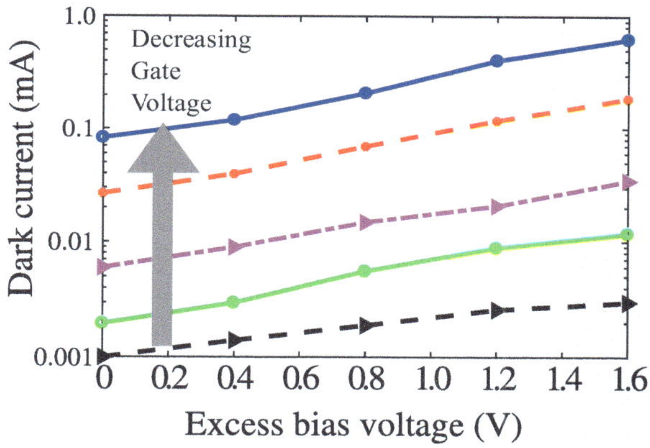

Fig. 6.5 pg-SPAD SiPM dark current as a function of excess bias voltage, V_{Exc}, for varying gate bias voltage, $|V_G|$

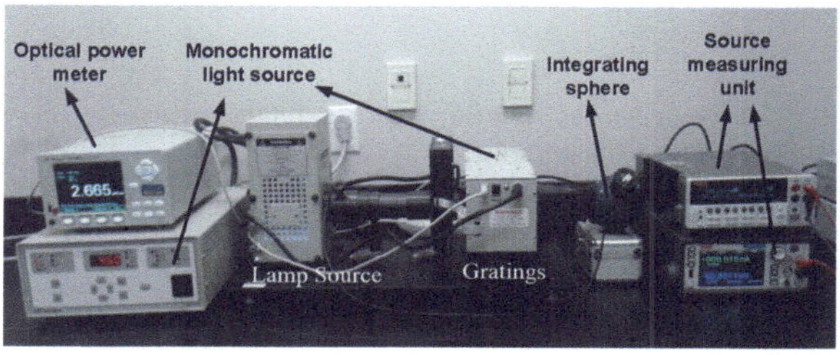

Fig. 6.6 Optical test bench for experiment in [14]

independently confirm the light levels, a monochromatic light source with ability to vary the wavelengths over a large range with nm resolution, an integrating sphere for generating uniform photon flux over the SiPM surface area, and source measure units for varying the required voltage biases. Optical couplers ensure the tunable monochromatic source is properly interfaced with the integrating sphere and detection surfaces. Monochromatic light sources may use a tungsten lamp with diffraction gratings to generate light of different wavelengths. Pulsed laser sources may also be used to generate the incident photons. Photon detection efficiency is one of the primary performance measure of SPADs and is calculated as

$$PDE = \frac{I_m hc}{P_{op} G \lambda e} \tag{6.1}$$

where I_m is the photon-generated current, h is Planck's constant, c is the speed of light, P_{op} is the optical power incident on the active area at a given wavelength (λ), G is the pixel gain, and e is the electron charge ($1.6 \times 10^{-}19$). The photon detection efficiency is intrinsically linked to the signal-to-noise ratio (SNR) of the SiPM.

The SiPM pixel gain can be written as

$$G = \frac{C V_{Exc}}{e} \tag{6.2}$$

where C is the capacitance and V_{Exc} is the excess bias voltage (amount of voltage being applied that is above the breakdown voltage).

The SNR of an SiPM is

$$SNR = \frac{I_m - I_d}{I_d} \tag{6.3}$$

where I_m is the photon-generated current and I_d is the current in the absence of photons (i.e., the dark current).

The pg-SPAD SiPM's spectral response as a function of gate and excess bias is shown in Figs. 6.7 and 6.8. Characterization measurements of commercially available SiPMs developed by companies such as Hamamatsu and SensL are characterized within the range of 10^8–10^{11} photons/s for a wavelength of 500 nm. Accordingly, in [14], the optical power was kept fixed at 100 nWcm^{-2} (2.5×10^{10} photons/s) to make it comparable with the commercially available SiPMs [20].

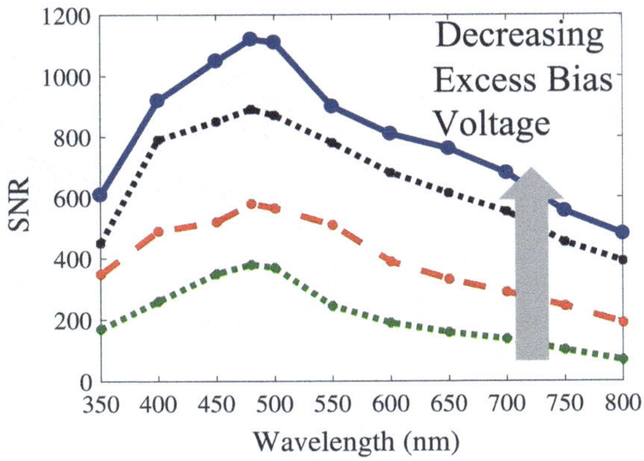

Fig. 6.7 Signal-to-noise ratio as a function wavelength for varying excess bias voltage ($|V_G| = 14$ V, optical power of 100 nWcm^{-2}) [14]

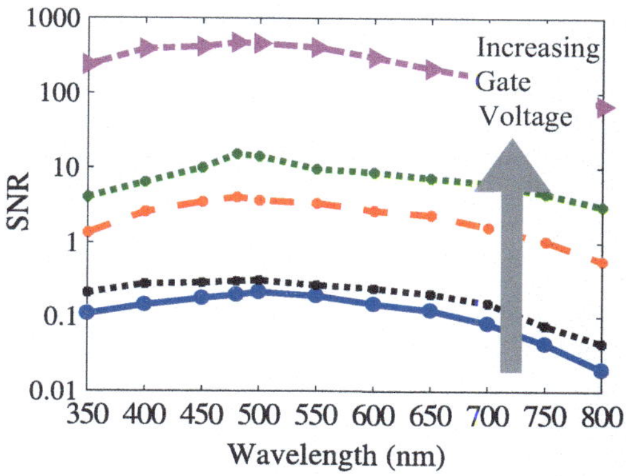

Fig. 6.8 Signal-to-noise ratio as a function of wavelength for varying gate voltage (excess bias of 0.4 V, optical power of 100 nWcm^{-2}) [14]

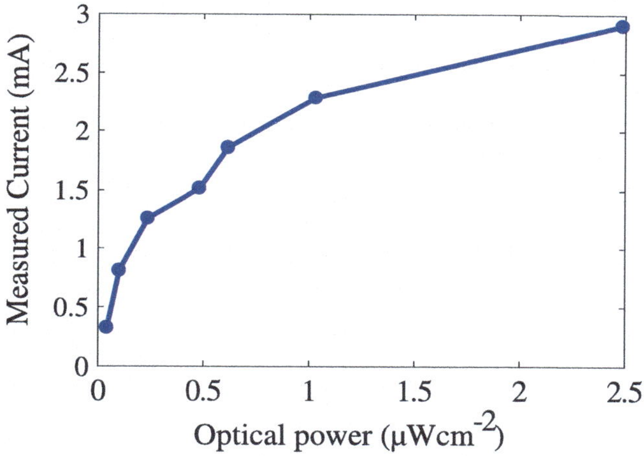

Fig. 6.9 Output current as a function of incident optical power (wavelength = 500 nm, gate voltage = 14 V, excess bias = 0.4 V) [14]

Increasing the excess bias increases the dark current and the sensitivity. However, these occur at different rates resulting in an overall decrease in the signal-to-noise ratio. For a fixed excess bias, the gate voltage has a significant effect on the signal-to-noise ratio as seen in Fig. 6.8 where the signal-to-noise ratio increases by a factor of 1000 as the absolute value of gate voltage is increased.

Figure 6.9 shows the output measured current as a function of optical power. The applied gate voltage magnitude was fixed at 14 V and an excess bias voltage of 0.4 V was used in [14]. A tunable monochromatic light with integrating sphere varied the optical power over the range of 0.042 to 2.486 μWcm^{-2} using an optical attenuator and optical filters. The wavelength was kept fixed at 500 nm during the measurements. The total output current increases with increasing photon flux (Fig. 6.9) resulting in a sensitivity of 1.06 kA per W/cm^2.

6.3.4 pg-SPAD SiPM: Temperature Characterization

Temperature affects the carrier generation process and the statistics of any trapping/detrapping process and tunneling processes. In [14], the noise performance of the designed pg-SPAD SiPM was characterized at three different temperatures, −60 °C, 25 °C, and 60 °C. These represent typical extremes that may be found in many applications. In temperature testing, a temperature chamber to hold the devices at a constant temperature while testing is needed. Many temperature chambers can heat up from room temperature as well as, with the help of liquid nitrogen or helium, go down to liquid nitrogen or liquid helium temperatures. Low-temperature measurements can also be obtained by freezing the devices in a commercial precision low-temperature freezer, such as those used for cell culture applications, and carefully monitoring the temperature as it rises to room temperature. In [14], while measuring, the gate voltage was kept constant, and for each temperature, two sets of data were taken with two excess bias voltages of 0.8 and 1.6 V. The result, shown in Fig. 6.10, confirms the increase in dark count with temperature due to an increase in the number of carriers increasing the probability of an avalanche. In fact many SiPMs are used when cooled to reduce the thermally generated dark events. Dark current increases with the excess bias. At higher temperatures, Shockley-Read-Hall generation dominated over band to band tunneling. The dark counts vary exponentially with temperature [21–23]. Thus, the dark current increases exponentially in the higher-temperature ranges. Band to band tunneling dominates at the lower temperatures and its correlation with temperature is relatively weak. This causes a slight increase at low temperature for the dark current [21–23, 31–33]. Due to the effect of the perimeter field gate, the pg-SPAD SiPM has a dark current that is more affected at lower temperatures.

The comparison of a pg-SPAD SiPM and other SiPMs found in the literature is summarized in Table 6.1 where the comparison is based on a single SiPM [16, 17,

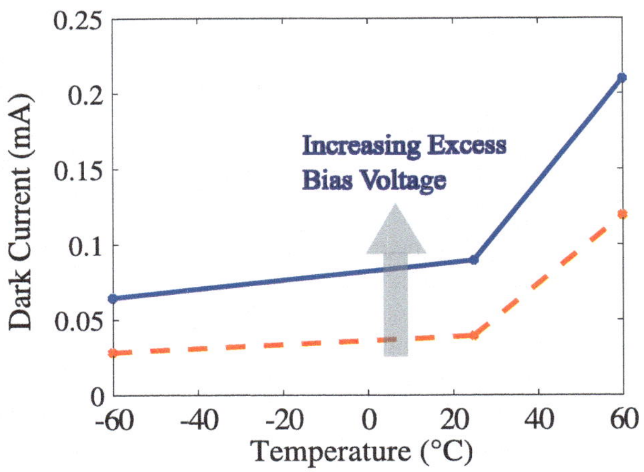

Fig. 6.10 Variation of dark current with temperature for pg-SPAD SiPM ($|V_G| = 5$ V) [14]

Table 6.1 Performance summary and comparison from [14]

Parameter	Our work	[17]	[15]	[16]	[18]	[19]	[24]
Process	0.5-μm standard CMOS	Specialized	0.13-μm CMOS imaging	0.35-μm high-voltage CMOS	High-volume CMOS	0.8-μm high-voltage CMOS (simulation only)	0.15-μm standard CMOS
Tech.	Analog	Analog	Digital	Multi-digital	Digital	Digital	Digital
Area (mm^2)	0.86 × 0.85	3.5 × 3.5	9.85 × 5.45	0.8 × 0.78	3.9 × 3.2	1.1 × 1.1	4.4 × 4.4
# SiPM	1	1	128	1	4	1	4096
Pixel size (μm^2)	33 × 33	2844.44	570 × 610	30 × 50	52 × 30	50 × 50	4726.56
Active area (μm^2)	121	1024	148120.2	318	780	1250	1252.54
Fill factor (%)	11.11	36	42.6	21.2	50	50	26.5
Tunable SNR	Yes	No	No	No	No	No	No
Tunable sensitivity	Yes	No	No	No	No	No	No

19] or a single pixel [15, 18, 24]. The pg-SPAD SiPM has a relatively lower fill factor than other implementations, and pg-SPAD SiPMs are the only option for tunable performance metrics while at the same time offering the benefits of integrating the readout electronics and increasing the speed. Note that in [14], the relatively low fill factor is due to the chosen pixel size. Increasing the overall pixel size does not increase the readout electronics, but will increase the active area, this generating a larger fill factor. This will also result in (generally) larger fill factors as lower-technology nodes (although analog transistors may still maintain similar sizing due to mitigating early voltage and short-channel effects). In [14], increasing the pixel size to $30 \times 50\,\mu\text{m}$ similar to the pixel size of [16] would create a pixel with a fill factor of $\approx 35\%$.

In summary, a CMOS SiPM developed solely with pg-SPADs and that is suitable for nuclear imaging applications has been used as a case study. The candidate SiPM is an 18×18 array with each pixel dimension on the order of $30\,\mu\text{m}$ and a fill factor of 11%. Larger fill factors can be obtained by increasing the pixel size, as while the optical area increases in tandem with the pixel size, the quenching circuits remain approximately the same. Simulations of CMOS SiPMs require a model that includes generating the DC voltage characteristics and the dynamic avalanche, quench, reset operation. The model also needs to include the carrier generation process and include the effect of a single photon. There are a number of models, described in an earlier chapter. The models are necessary as, particularly for CMOS implementations, the design cycle can be on the order of 6 months to a year depending on the foundry. As such, the ability to design and simulate before fabrication is invaluable to a designer. Shawkat et al. [14] developed the first noise model of the pg-SPAD and fully characterized the device for performance metrics including the dark current, signal-to-noise ratio, sensitivity, and temperature and the model was validated over the visible spectral range.

6.4 Dead Time Reduction Strategies

Dead time is the length of time during which a SPAD is nonresponsive to incoming photons. The simplified explanation is that once a photon has triggered an avalanche, a second photon my indeed trigger another avalanche; however, this is indistinguishable from the signal due the first photon. Thus, the avalanche needs to be completely quenched before the device is ready to "see" another photon. This places a limit on the maximum count rate of a SPAD and the maximum photon flux that can be detected. For active quenching, the measured count rate of the SPAD is [5, 6]

$$MCR = \frac{TCR}{1 + TCR \times t_{deadtime}} \tag{6.4}$$

where TCR is the true count rate of the device. At high light levels (or photon flux), the count rate will be maximum at $1/t_{deadtime}$. For passive quenching, the count rate is

$$MCR = \frac{TCR \times e^{-TCR \times t_{deadtime}}}{1 + TCR \times t_{deadtime}} \tag{6.5}$$

where the dead time takes into account the fact that an incident photon impinging on the device during the avalanche process is indistinguishable from the original photon.

While this may not be an issue for low-light-level applications such as fluorescence imaging or topologies such as quanta imagers for ultrahigh-resolution, this could be an issue when imaging in bright light. The dead time is a function of the parasitics of the device and reflects the time for the carriers to reset. Given the variation in parameters of different technology nodes, it is important to understand how the technology node affects the device. In general, dead time is reduced as the technology node decreases [8]. It should be noted that many readout topologies also focus on reducing the effects of afterpulsing. Thus, using the dead time as a figure of merit for a given device with an active readout can render comparison with other devices difficult.

The mitigation strategies to reduce dead time include layout techniques and circuit topologies. We review some of these techniques below.

6.4.1 Circuit Techniques for Dead Time Reduction

One technique for reducing the dead time is related to silicon photomultipliers. Instead of having a single large SPAD pixel, instead create smaller sub-pixels whose output is then summed to provide the overall output (Fig. 6.11). While the dead time is reduced to size reduction of the device, i.e., reducing the parasitic resistance and

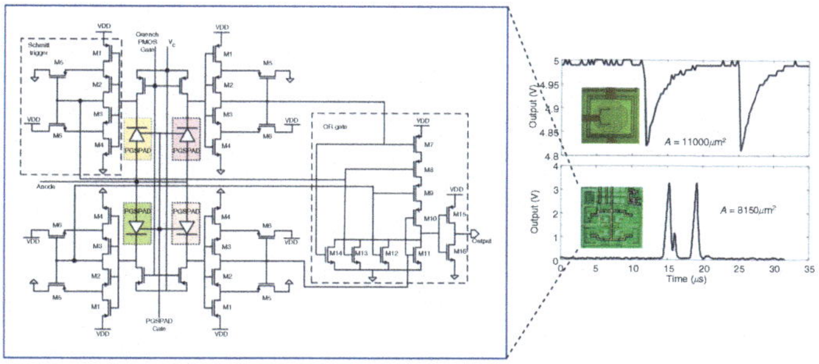

Fig. 6.11 Dead time reduction using subpixels instead of a single large pixel

capacitance, there is a penalty in terms of having more circuitry inside the pixel. Another more popular technique is to keep the pixel size constant and develop fast quenching circuits. Dead times ranging from ns to ps have been reported [4, 7, 12]. Dual-threshold quenching circuits make use of the parasitic capacitance to control the hold-off time (Fig. 6.11) [3].

Active quenching with monostable readout can also be used [4]. For applications such as quantum key distribution, dead time can represent a security concern [9, 10]. For optical receiver applications it has been theoretically demonstrated that increasing SPAD dead time increases the bit error rate particularly at high rates where counts are missed [11]. This is done by using a Poisson process to model the probability of detecting n photons and considering the counting statistics of the detector with both a modulated and unmodulated signal. As with the 4-sub-pixel detector, these missed counts can be mitigated by using an array of SPADs where the count rate of all the SPADs is summed. In the 4-sub-pixel array, a digital architecture is used. However, an analog architecture where all the avalanche currents are summed can also be used.

Afterpulsing, due to trapped charges that are subsequently released, is typically limited by ensuring the applied bias across the SPAD remains below breakdown. Afterpulsing is a function of the device itself, the excess bias, and the intrinsic capacitances present. Thus, there is some trade-off in circuit topology for maximizing the count rate and limiting afterpulsing. In many schemes, a mixed active/passive quenching approach along with a multiplexed outputs to achieve low afterpulsing and low dead time can be used [12, 13].

6.4.2 Reduced Dead Time Pixel

SiPMs have many devices in parallel, each of which can avalanche and quench independently. This idea can be used to build a pixel with reduced dead time. If n sub-pixels are used instead of a single large pixel, each can avalanche and quench independently. Dead time is a function of the area of the SPAD and is related to the capacitance of the device. Thus, a smaller device should have a smaller capacitance and thus a shorter dead time. In Ref. [1], four identical pg-SPADs are used as the sub-pixels. Each SPAD can avalanche, and while it is in the dead time of the device, one of the three other SPADs can still respond to incoming photons. As before, the quenching resistor is implemented as a transistor operating in triode to allow for a variable quenching resistance value. Instead of summing the avalanche current output of each pixel, this reduced dead time pixel operates as a small digital SiPM. The avalanche spike can be sent to a comparator or Schmitt trigger to digitize the output. A Schmitt trigger takes less space on the chip, leading to a better fill factor than a standard high-speed comparator. Opamps are not recommended as comparators as the slew rate is to low for most SPAD devices. The digitized output from the four SPADs can then be summed through an OR gate. The architecture, first developed in Ref. [1], is shown in Fig. 6.11 where all the anodes of the device are

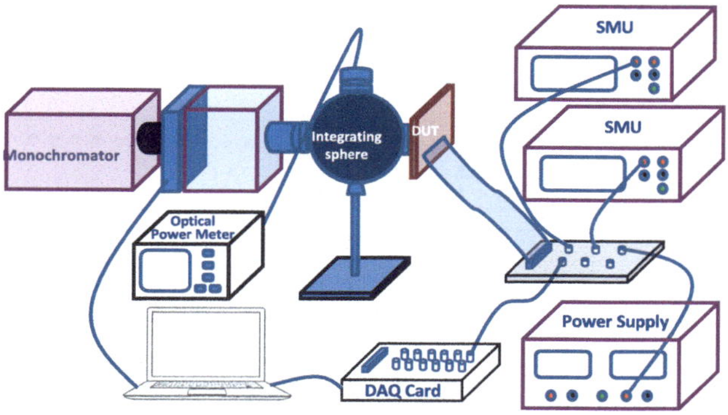

Fig. 6.12 A typical test setup will use source measure units to sweep the voltages and measure current as well as light sources, optical power meters, integrating sphere, and digitization and control. This setup can be used for I-V characterization without the optical equipment [1]

shorted together. The Schmitt triggers are the standard six-transistor configuration and are all biased in the same manner. A standard CMOS OR gate with eight inputs is used to sum the outputs.

Figure 6.12 is based on source measure units for applying the common gate and anode biases. Care should be taken to limit the current at these nodes to prevent large current surges damaging the device. It should be noted that the gate voltage bias does not have any significant effect on the forward bias but does have significant effect on the reverse bias. Figure 6.15 shows the I-V characteristics of the detector. SPICE simulations using the pg-SPAD SPICE model are shown in Fig. 6.13 [1]. The upper bound for the reduction in dead time for every n pixels follows an inverse law with a reduction in dead time by $1/n \times 100\%$. Four sub-pixels showed an improvement of roughly 25% in dead time.

Figures 6.14, 6.15, and 6.16 show the results of this dead time reduction technique. This technique does have limitations as it depends on the actual dead time of the device which is a function of the time constants associated with the device and readout circuits. The OR gate output is dependent on the width of the Schmitt trigger output and is a function of the switching points as well as the SPAD characteristics. These are both a function of sizing and area. As the number of sub-pixels increases, the pulse width sets a fundamental limit on the percentage decrease in dead time minimization that can be achieved.

The architecture of Fig. 6.11 was used to develop a 3×3 digital SiPM in [2]. In this form, each of the gate voltages can be individually controlled, and the noise of all the pixels can be compensated for. While the gate inputs can be controlled with electronics, an easy first pass used a data acquisition board and simple code to equalize the dark count rate across the small array. The noise was reduced with the gate voltage and the variation in the noise was easily compensated for to equalize the histogram of the array (Figs. 6.17 and 6.18).

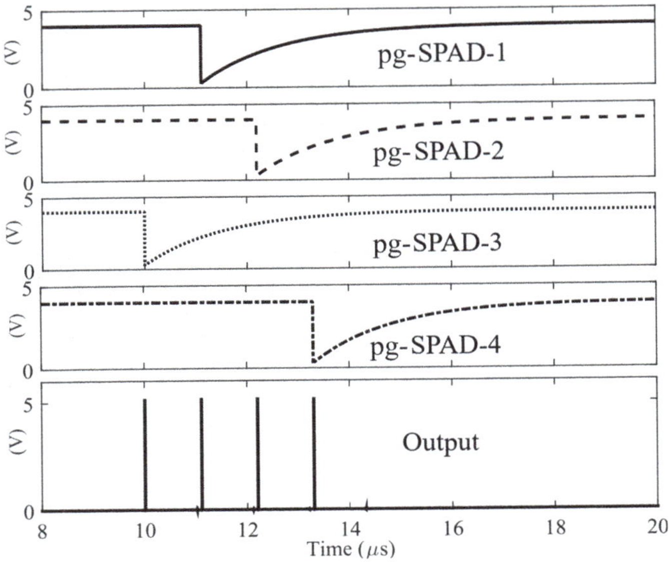

Fig. 6.13 A single pixel can be nonresponsive, while the other pixels are ready to respond to incoming photons [1, 2]

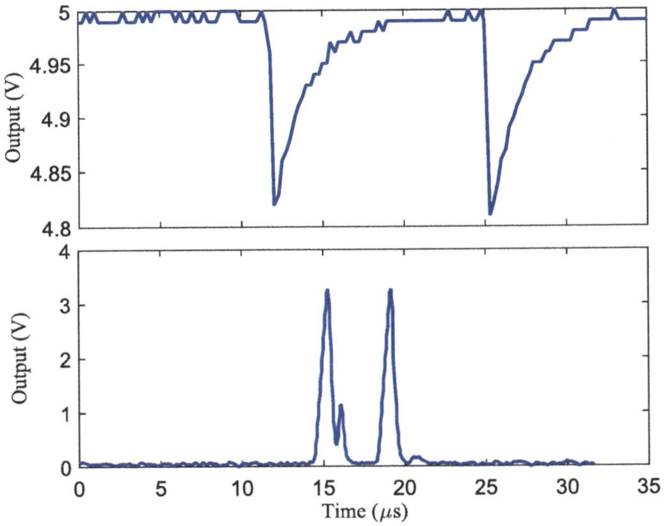

Fig. 6.14 Experimental results showing the dead time minimization in [1]

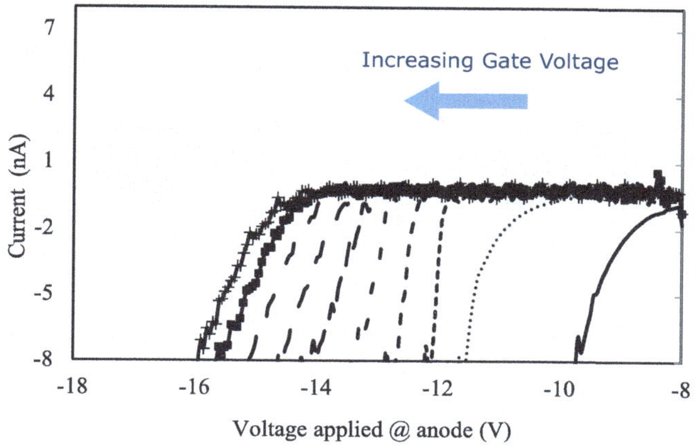

Fig. 6.15 I-V characteristics showing the gate voltage improving the breakdown voltage in [1]

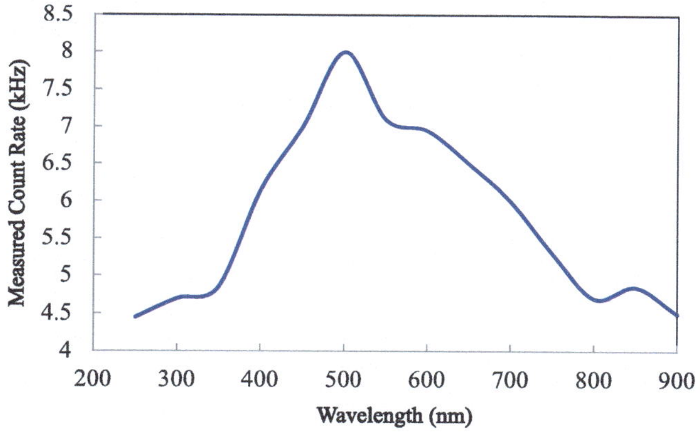

Fig. 6.16 Spectral response in [1]

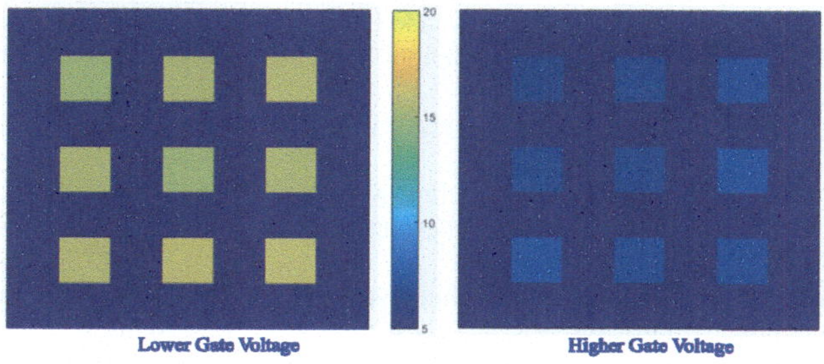

Fig. 6.17 Color map showing the dark count rate of an array of digital SiPM noise as a function of gate voltage. From left to right, the gate voltage increases

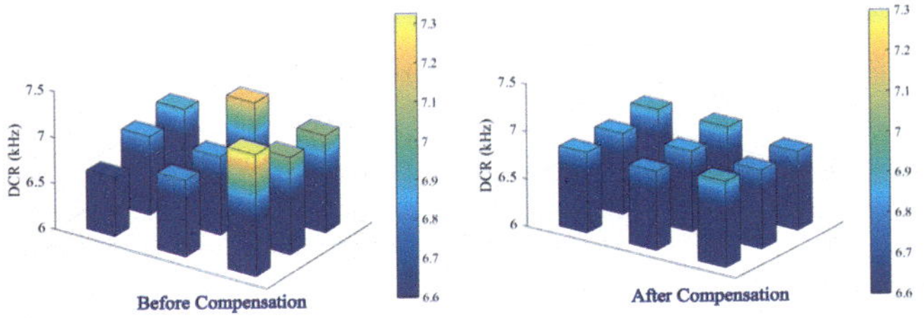

Fig. 6.18 The variation in dark count rate can be compensated for by applying unique gate voltages to each device [1]

6.5 Conclusions

In this chapter, we presented the use of pg-SPADs in SiPM architectures. We presented optoelectronic characterization results from a fabricated prototype. Furthermore, in addition to conventional circuit techniques, we offered an SiPM approach to deploy parallel pg-SPAD units per pixel in order to minimize dead time.

References

1. M.H.U. Habib, M.S.A. Shawkat, N. McFarlane, A tunable single photon avalanche diode pixel with improved time resolution, in *IEEE Sensors*, Glasgow (2017), pp. 1–3. https://doi.org/10.1109/ICSENS.2017.8234050
2. M.H.U. Habib, M.S.A. Shawkat, N. McFarlane, Improved signal to noise ratio across the spectral range for CMOS silicon photomultipliers, in *2016 IEEE Sensors*, Orlando (2016), pp. 1–3. https://doi.org/10.1109/ICSENS.2016.7808588
3. C. Niclass, M. Soga, A miniature actively recharged single-photon detector free of afterpulsing effects with 6ns dead time in a 0.18 μm CMOS technology, in *2010 International Electron Devices Meeting*, San Francisco (2010), pp. 14.3.1–14.3.4. https://doi.org/10.1109/IEDM.2010.5703360
4. A. Eisele, R. Henderson, B. Schmidtke, T. Funk, L. Grant, J. Richardson, W. Freude, 185 MHz count rate, 139 dB dynamic range single-photon avalanche diode with active quenching circuit in 130 nm CMOS technology
5. S. Lee, R. Gardner, A new G-M counter dead time model. Appl. Radiat. Isot. **53**, 731–737 (2000)
6. L. Neri, F. Musumeci, A. Scordino, G. Fallica, M. Mazzillo, M. Zimbone, Dead time causes and correction method for single photon avalanche diode devices. Rev. Sci. Instrum. **81**, 086102 (2010)
7. O. Kumagai et al., 7.3 A 189×600 back-illuminated stacked SPAD direct time-of-flight depth sensor for automotive LiDAR systems, in *2021 IEEE International Solid-State Circuits Conference (ISSCC)*, San Francisco (2021), pp. 110–112. https://doi.org/10.1109/ISSCC42613.2021.9365961
8. X. Qian, W. Jiang, M.J. Deen, Single photon detectors for automotive LiDAR applications: state-of-the-art and research challenges. IEEE J. Sel. Top. Quantum Electron. **30**(1), 1–20 (2024). Single-Photon Technologies and Applications, Art no. 3800520. https://doi.org/10.1109/JSTQE.2023.3304294

9. S. Jiang, M. Safari, High-speed free-space QKD in the presence of SPAD dead time, in *2022 IEEE International Conference on Communications Workshops (ICC Workshops)*, Seoul, Korea (2022), pp. 457–462. https://doi.org/10.1109/ICCWorkshops53468.2022.9814553
10. V. Burenkov, B. Qi, B. Fortescue, H.-K. Lo, Security of high speed quantum key distribution with finite detector dead time. Quantum Info. Comput. **14**(3–4), 217–235 (2014)
11. E. Sarbazi, M. Safari, H. Haas, The impact of long dead time on the photocount distribution of SPAD receivers, in *2018 IEEE Global Communications Conference (GLOBECOM)*, Abu Dhabi (2018), pp. 1–6. https://doi.org/10.1109/GLOCOM.2018.8647814
12. F. Severini, I. Cusini, D. Berretta, K. Pasquinelli, A. Incoronato, F. Villa, SPAD pixel with sub-NS dead-time for high-count rate applications. IEEE J. Sel. Top. Quantum Electron. **28**(2), 1–8 (2022). Optical Detectors, Art no. 3802808. https://doi.org/10.1109/JSTQE.2021.3124825
13. P. Keshavarzian et al., A 3.3-Gb/s SPAD-based quantum random number generator. IEEE J. Solid-State Circuits **58**(9), 2632–2647 (2023). https://doi.org/10.1109/JSSC.2023.3274692
14. M.S.A. Shawkat, M.H.U. Habib, N. McFarlane, An analog CMOS silicon photomultiplier using perimeter gated single photon avalanche diodes. IEEE Trans. Circuits Syst. I (TCAS-I) **65**(11), 3830–3841 (2018)
15. L.H. Braga, L. Gasparini, L. Grant, R.K. Henderson, N. Massari, M. Perenzoni, D. Stoppa, R. Walker, A fully digital 8 $\times$ 16 SiPM array for pet applications with per-pixel TDCs and real-time energy output. IEEE J. Solid-State Circuits **49**(1), 301–314 (2014)
16. S. Mandai, V. Jain, E. Charbon, A 780 $\times$ 800 μm^2 multichannel digital silicon photomultiplier with column-parallel time-to-digital converter and basic characterization. IEEE Trans. Nucl. Sci. **61**(1), 44–52 (2014)
17. M. Mazzillo, G. Condorelli, D. Sanfilippo, G. Valvo, B. Carbone, A. Piana, G. Fallica, A. Ronzhin, M. Demarteau, S. Los, E. Ramberg, Timing performances of large area silicon photomultipliers fabricated at STMicroelectronics. IEEE Trans. Nucl. Sci. **57**(4), 2273–2279 (2010)
18. T. Frach, G. Prescher, C. Degenhardt, R. de Gruyter, A. Schmitz, R. Ballizany, The digital silicon photomultiplier—principle of operation and intrinsic detector performance, in *IEEE Nuclear Science Symposium Conference Record* (2009), pp. 1959–1965
19. M.-A. Tétrault, A.C. Therrien, W. Lemaire, R. Fontaine, J.-F. Pratte, TDC array tradeoffs in current and upcoming digital SiPM detectors for time-of-flight PET. IEEE Trans. Nucl. Sci. **64**(3), 925–932 (2017)
20. R. Bencardino, J.E. Eberhardt, Development of a fast-neutron detector with silicon photomultiplier readout. IEEE Trans. Nucl. Sci. **56**(3), 1129–1134 (2009)
21. E. Vilella, O. Alonso, A. Montiel, A. Diéguez, Characterization of temperature effects on single-photon avalanche diodes fabricated in a HV-CMOS conventional technology, in *IEEE Symposium on Design, Test, Integration and Packaging of MEMS/MOEMS* (2013), pp. 1–5
22. D.A. Ramirez, M.M. Hayat, M.A. Itzler, Dependence of the performance of single photon avalanche diodes on the multiplication region width. IEEE J. Quantum Electron. **44**(12), 1188–1195 (2008)
23. D. Massey, J. David, G. Rees, Temperature dependence of impact ionization in submicrometer silicon devices. IEEE Trans. Electron Devices **53**(9), 2328–2334 (2006)
24. M. Perenzoni, D. Perenzoni, D. Stoppa, A 64$\times$64-pixels digital silicon photomultiplier direct TOF sensor with 100-MPhotons/s/pixel background rejection and imaging/altimeter mode with 0.14% precision up to 6 km for spacecraft navigation and landing. IEEE J. Solid-State Circuits **52**(1), 151–160 (2017)
25. A.D. Guerra, N. Belcari, M.G. Bisogni, G. Llosá, S. Marcatili, S. Moehrs, Advances in position-sensitive photodetectors for PET applications. Nucl. Instrum. Methods Phys. Res. Sect. A: Accel. Spectrom. Detect. Assoc. Equip. **604**(1–2), 319–322 (2009)
26. P. García de Acilu, P. Mendes, M. Canadas, Evaluation of APD and SiPM matrices as sensors for monolithic PET detector blocks, in *2011 IEEE Nuclear Science Symposium Conference Record*, Valencia (2011), pp. 3300–3304. https://doi.org/10.1109/NSSMIC.2011.6152594

27. L.H.C. Braga, L. Gasparini, L. Grant, A fully digital 8 × 16 SiPM array for PET applications with per-pixel TDCs and real-time energy output. IEEE J. Solid-State Circuits **49**(1), 301–314 (2014). https://doi.org/10.1109/JSSC.2013.2284351
28. M.H.U. Habib, N. McFarlane, A perimeter gated single photon avalanche diode based silicon photomultiplier as optical detector, in *2015 IEEE 58th International Midwest Symposium on Circuits and Systems (MWSCAS)*, Fort Collins (2015), pp. 1–4. https://doi.org/10.1109/MWSCAS.2015.7282084
29. X. Liang, N. D'ascenzo, W. Brockherde, S. Dreiner, A. Schmidt, Q. Xie, Silicon photomultipliers with area up to 9 mm^2 in a 0.35-μ m CMOS process. IEEE J. Electron Devices Soc. **7**, 239–251 (2019). https://doi.org/10.1109/JEDS.2019.2893802
30. J. Preitnacher, S. Ageev, W. Hansch, A novel segmented SiPM combined with amplifiers on one chip with improved single-photon timing resolution in 0.35-m CMOS. IEEE Trans. Nucl. Sci. **71**(3), 301–308 (2024). https://doi.org/10.1109/TNS.2024.3363228
31. A. Buchner, S. Hadrath, R. Burkard, F.M. Kolb, J. Ruskowski, M. Ligges, A. Grabmaier, Analytical evaluation of signal-to-noise ratios for avalanche- and single-photon avalanche diodes. Sensors **21**, 2887 (2021). https://doi.org/10.3390/s21082887
32. H. Mahmoudi, M. Hofbauer, B. Goll, H. Zimmermann, Noise and breakdown characterization of SPAD detectors with time-gated photon-counting operation. Sensors (Basel). **21**(16), 5287 (2021). https://doi.org/10.3390/s21165287. PMID: 34450728; PMCID: PMC8400203.
33. Y. Xu, P. Xiang, X. Xie, Comprehensive understanding of dark count mechanisms of single-photon avalanche diodes fabricated in deep sub-micron CMOS technologies. Solid-State Electron. **129**, 168–174 (2017). ISSN 0038-1101. https://doi.org/10.1016/j.sse.2016.11.009

Chapter 7
Readout Strategies and Asynchronous Architectures

7.1 Introduction

Silicon photomultipliers are widely used in a number of different applications in the health, environmental, and scientific discovery fields. Silicon photomultipliers (SiPMs) are semiconductor-based optical detectors that, when implemented in silicon-based semiconductors, image primarily in the visible region of light. Silicon photomultipliers are not significantly affected by magnetic fields and have lower operating voltages compared to photomultiplier tubes (PMTs). SiPM properties are also relatively consistent over time, and aging does not affect the SiPM's performance [2–4]. In addition to this, SiPMs implemented in commercial CMOS foundries allow for the inclusion of the electronics directly on chip, leading to reduced parasitics and higher throughput and speed. Due to this, SiPMs are being widely used to replace PMTs and other photodetectors in applications such as light detection and ranging (LiDAR) [5–8], optical imaging systems [9–13], medical imaging [14–18], fluorescence imaging [19–21], astrophysics, and gamma detection [22–31] where they may be used in different types of experiments such as time-correlated single-photon counting [32–34].

An SiPM is an array of single-photon avalanche diodes, or SPADs, and quenching devices. The overall structure is shown in Fig. 7.1a. Each SPAD is biased in reverse breakdown (Geiger mode), with a quenching device in series with it. When photons are incident on the SPADs, electron–hole pairs are created and accelerated under the high electric field. They undergo impact ionization leading to avalanche multiplication and a large current. The quenching device can be implemented using passive devices (resistor) or active devices (MOSFET). A simple SiPM-equivalent circuit model is shown in Fig. 7.1b, where the model is divided into avalanching vs non-avalanching cells, and each cell consists of parasitic devices [35–37]. The model shows only a single avalanching cell. To account for more cells avalanching at a given time, the model would be adjusted to have F firing microcells and M non-avalanching microcells where $F + M = N$.

M. Dandin et al., *Single-Photon Avalanche Diodes and Photon Counting Systems*,
https://doi.org/10.1007/978-3-031-64334-7_7

It should be noted that all cells do not necessarily fire at the same time, and that there may be a small delay that is not captured in this model. This model also does not take into account the full probabilistic nature of the photon-induced avalanche process, but is suitable for designing, simulating, and characterizing the readout electronics' performance before committing to potentially expensive devices or fabrication. The parasitic capacitances are due to the parallel and fringe capacitances of the metal interconnects and the depletion capacitances of the diodes. The use of devices integrated on the same chip offers lowered parasitics compared to off-chip devices. SiPMs can be either analog or digital [1]. In analog SiPMs, the output is measured at the common cathode or the common resistor node or through a capacitor connected at the resistor–SPAD junction (Fig. 7.1a). The other end of all the capacitors is a common node V_o. This approach offers improved timing performance. In digital SiPMs, the output of each SPAD is taken directly at the node between the quenching resistor and diode through its own dedicated readout circuit. In this case, the output of the SPAD is typically a digital signal. SiPMs have been implemented using custom fabrication processes and commercial foundry processes. Implementing SiPMs in commercial foundries can require an initial fabrication run to develop test devices in order to extract parameters suitable for simulating the model shown in Fig. 7.1b with the designed electronics. However, in recent years, a few commercial foundries, such as XFAB, have begun offering single-photon avalanche diodes as part of their fabrication options which will enable easier simulations of the SiPMs and the electronics on the same chip.

SiPM-based imaging systems can consist of commercially available SiPMs with discrete off-the-shelf or custom electronics on a PCB board [38]. However, this approach generally has higher parasitics which may limit the system speed. It also has optical dead zones which decrease the fill factor and spatial resolution. Integration of the electronics in a single chip can reduce those parasitics. However, the electronics, particularly if complex with multiple stages, could potentially use the available area for the optical detector, reducing the overall fill factor across the

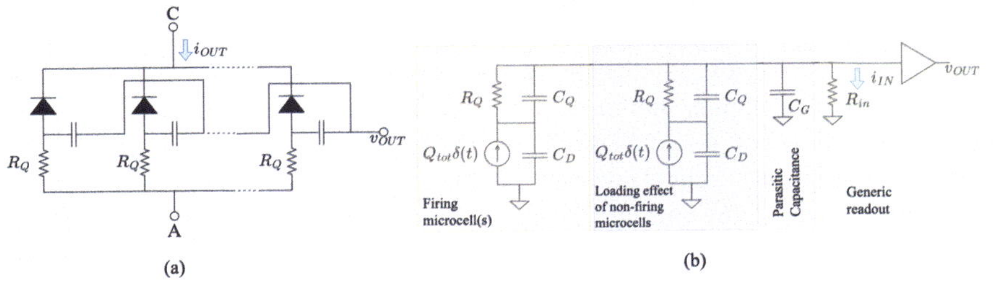

Fig. 7.1 (**a**) An SiPM is an array of SPADs each with its own quenching resistor R_Q. For analog SiPMs the output is the sum of the avalanche currents taken at either the common cathode C or common anode A. (**b**) SiPM circuit model showing 1 avalanching cell and (N-1) non-avalanching cells. C_Q is the quenching device's parasitic capacitance, C_D is the diode parasitic capacitance, and C_G is the equivalent parasitics associated with routing. R_{in} represents the input impedance of the readout circuit [37]

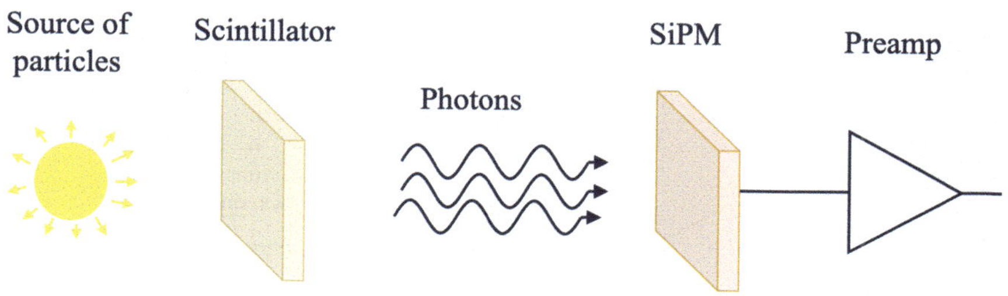

Fig. 7.2 Scintillators convert nuclear particles into photons which are then converted by the SiPM into electrons. These can be amplified and further processed using charge-, current-, or voltage-based readout electronics

chip. Fill factor is typically calculated as the ratio of the optical active area, typically defined by the depletion region of the pn junction, and the area of non-optical-related devices such as resistors and transistors, as well as any metal routing combined with the total pn device area. This integration of readout and optical devices on the same chip can be accomplished by implementing both the SiPM and electronics in standard CMOS processes [39, 40]. In vertical integration, the diodes are in one layer, while the other devices are in a separate layer. Thus, this offers the largest fill factor compared to a 2D implementation [41, 42].

In this review, the primary focus is on nuclear imaging which is accomplished using a scintillator material, such as LYSO (lutetium–yttrium oxyorthosilicate), coupled to the SiPM. Scintillators emit light when particles, such as gamma rays or neutrons, are incident on them. A schematic of a typical nuclear imaging system is shown in Fig. 7.2, where time of arrival, energy, and position are the three quantities of interest. There are a number of options for integrating the SiPM and scintillator. The exact method will depend on the material used to fabricate the scintillator. The method of integrating the two will also depend on the application, as this will drive the accuracy, sensitivity, and resolution constraints of the entire system. In general, the SiPM and scintillator can be integrated using commercially available optical coupling gels or grease. More details on the performance parameters of optical coupling materials can be found in [43]. A typical readout architecture, whether in integrated circuits or discrete off-the-shelf components, is composed of a number of modules and is shown in Fig. 7.3.

Regardless of the application, the initial SiPM output signal must be amplified before being further processed and digitized. The three typical approaches for front-end amplification that are used to readout the SiPM arrays consist of charge, transimpedance, or voltage amplification and are shown in Fig. 7.4. The chapter first gives an overview of the preamplifier structures. This is followed by a discussion on energy, timing, and position measurement along with a summary of array readout strategies. Finally, our conclusion summarizes the chapter and describes the challenges and possible future paths.

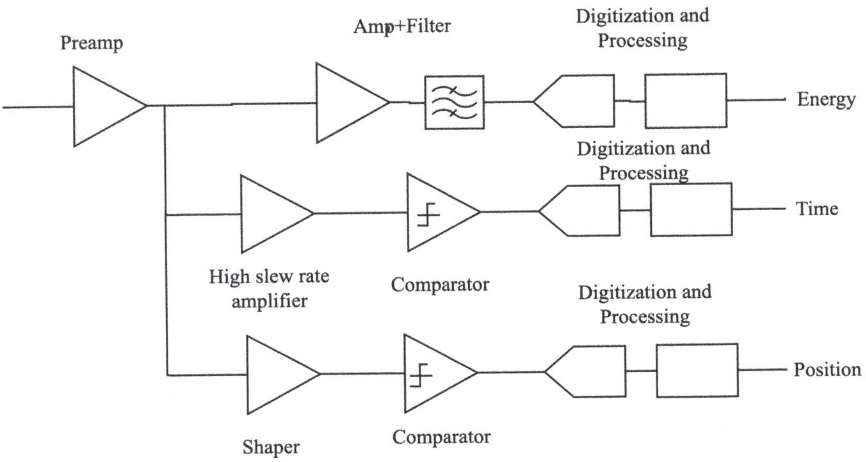

Fig. 7.3 The typical readout in a nuclear imaging system requires various circuit topologies to extract time, energy, or position information

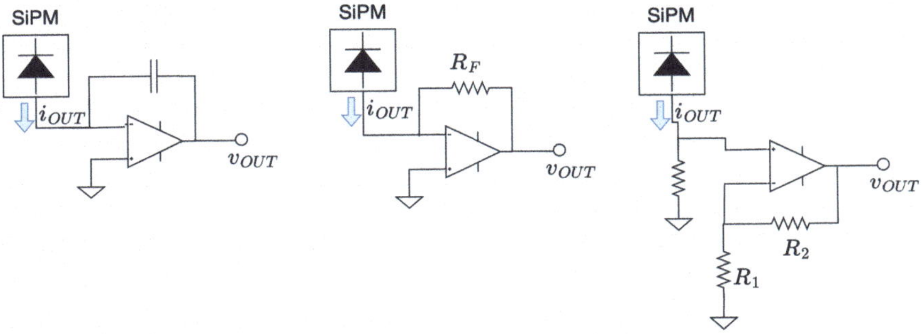

Fig. 7.4 Typical preamplifier topologies: (**a**) charge amplifier, (**b**) transimpedance amplifier, and (**c**) voltage amplifier

7.2 Preamplifier Structure in Nuclear Imaging Systems

The preamplifier is the front-end circuitry between the raw SiPM readout and the signal processing electronics. Thus, its characteristics directly affect the performance of subsequent stages. The preamplifier stage can be implemented as either single ended or fully differential. When implemented as a fully differential amplifier, a common-mode feedback stage is required to maintain the common mode at a known level [44]. Typical common-mode approaches will take the difference of the average of the amplifier outputs with the common-mode voltage to adjust the tail current or the currents in the active loads to move the common-mode voltage up or down as required. Implementations of common-mode feedback can use either diff amp-based or switch capacitor-based methods. Reference [45] describes an auxiliary amplifier-based common-mode feedback circuit (CMFB) that has a phase margin of 62.1° and gain margin of 25 dB.

7.2.1 Charge-Sensitive Preamplifiers (CSPs)

A charge-sensitive preamplifier has an output voltage that is proportional to the integrated input current. A typical architecture consists of an operational amplifier, a capacitor in the feedback path, and a reset switch or circuit. The reset circuit discharges the feedback capacitor; otherwise, the output voltage eventually saturates. High input impedance guarantees the current flows primarily through the feedback capacitor. Due to the Miller effect, the feedback capacitor (C_f) appears as large capacitance at the input ($C_{eff} = (1 + A)C_f$) which is in parallel with other parasitic input capacitances. Thus, the gain of this structure is proportional to the feedback capacitance and is relatively independent of the devices' capacitances. The feedback network can be implemented using passive or active components. The advantage of active feedback (transistors) over passive networks (RC) lies in the active devices' lower noise and tunability. In addition, a small chip area is required to implement the circuit. An active feedback with leakage current compensation is highlighted in [46]. Reference [47] uses an active feedback network for gamma-ray tracking detectors with a low-gain, noninverting amplifier between the charge amplifier and the feedback resistance. Thus, the discharge time constant can be made shorter while leaving the RC network unchanged. In another work, an active feedback network is designed upon two different active feedbacks based on the MOSFET devices and a voltage-controlled switch as the reset network. Using this technique, the device can operate at as high as 4.5 MHz and can readout 1 to 10^4 photons with an energy of 1 and 10 keV [48]. In [49], a two amplifier feedback network replaces the large resistor. One amplifier maintains the DC offset voltage at zero and the other one constitutes the negative feedback around the main amplifier.

The opamps used for CSPs can be based on BJT transistors [50, 51] or MOSFET and JFET transistors [52–54]. JFET- and BJT-based opamps are widely available commercially. In comparison to MOSFET, JFET CSPs provide better noise performance. A noise model for JFET CSPs can be found in [52]. The result of this work revealed that, at 10 kHz, the JFET amplifier's input-referred inherent noise exhibited 22-dB improvement as opposed to MOS-based CSPs.

However, MOSFET-based approaches can offer reduced size and the ability to integrate the SiPM with the amplifier on the same chip [44, 55–58]. BiCMOS (bipolar and CMOS on the same die) technologies are less radiation-hard compared to CMOS [58], and in general CMOS is less radiation tolerant than SOI (silicon on insulator) [59, 60]. The irradiation tests in [58] demonstrate that, with high-input gamma rays, the CSPs' output amplitude signal and the SNR reduces up to 34.3% and 11.6 dB. Moreover, the fall time also increases significantly from 201 to 1730 ns. This research also indicates that MOS transistors are more tolerant to radiation sources and provide better performance than BiCMOS.

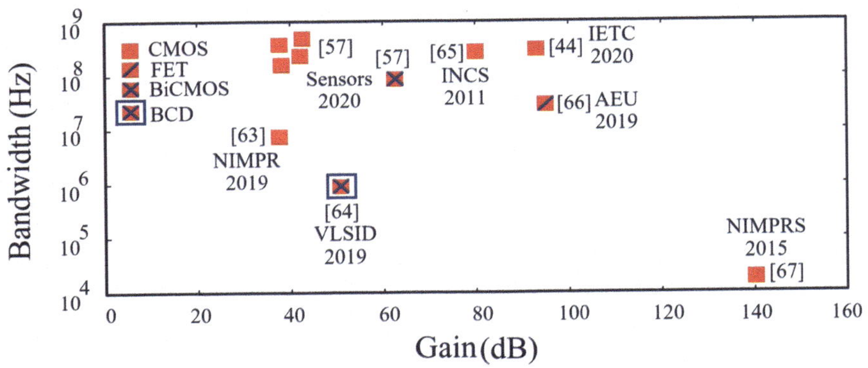

Fig. 7.5 Gain and bandwidth of charge amplifiers [45, 58, 64–68]

A challenge for CMOS-based CSPs is the relatively large area capacitor. A typical capacitor may be 10 times as large as typical analog-sized transistor. The capacitor has to discharge periodically. This can be performed using feedback resistors, transistors, and other novel active devices [44, 46–48]. Passive resistance is common; however, the large value has implications on the opamp DC characteristics and requires large area if implemented in integrated circuits. R_f can be implemented with MOS transistors to reduce area and provide greater design flexibility [61]. The offset, bandwidth, power consumption, dynamic range, and gain of CSP are functions of the feedback components [45, 49, 62, 63]. Figure 7.5 shows a comparison of the gain and bandwidth characteristics of a few implemented charge amplifiers.

7.2.2 *Transimpedance Preamplifier*

The conventional preamplifier used in nuclear imaging systems comprises a charge-sensitive and shaping amplifier [69]. An ideal nuclear imaging system needs to be fast, with a small rise time and dynamic slew rate correction and be low power and small sized for radiation detection readout [70, 71]. However, even with design techniques, the CSP's slew rate is in the microsecond range [72]. In practice, CSP and the shaping amplifier can be substituted with a transimpedance amplifier, where the current can be directly converted into an output voltage [73].

In general, a typical TIA implementation using off-the-shelf discrete components is a feedback resistor with an amplifier. For large gain, the input resistance, $R_{in}=R_f/(A+1)$, is minimized. The TIA input slew rate is a function of the input RC time constant. Lower RC gives a higher slew rate and is suitable particularly for high count rate applications [69, 73–78]. The input impedance of a transimpedance amplifier is

$$Z_{in} = \frac{R_f}{g_m R_o} + \frac{sR_f C_o}{g_m} = \frac{R_f}{g_m}\left(\frac{1}{R_o} + sC_o\right) \tag{7.1}$$

where R_f is the resistor in the feedback path, g_m is the transconductance gain, C_0 is the internal capacitance of amplifier, and R_0 is load resistance. This can cause oscillations when coupled with the large input capacitance from an SiPM. To mitigate this, a capacitance is placed in parallel with the feedback resistor. The large feedback resistor may be implemented by an MOS device operating in triode [79, 80]. In order to detect and transform nanosecond-wide signals, the TIA requires a large bandwidth. A TIA's bandwidth can be written as

$$BW_{TIA} = \sqrt{\frac{GBW}{2\pi R_F C_D}} \tag{7.2}$$

where BW_{TIA} is the bandwidth, R_F is the resistance, GBW is the gain bandwidth product, and C_D is the detector capacitance. The GBW is a constant for the open-loop opamp or OTA structure. In [72], in order to increase the bandwidth, two inductors are in series with the TIA. There are multiple approaches to implement a suitable TIA for nuclear imaging [81]. Reference [75] proposes a TIA based on second-generation voltage conveyors (VCII). This voltage conveyer has been implemented using discrete components [76, 77]. VCII, compared to the first generation, provides variable front-end gain and low impedance at the input of VCII that helps to decrease the effect of usually large output parasitic capacitance associated with SiPM.

Common base and common gate amplifiers provide low input impedance and do not require feedback resistances. These single transistor amplifier architectures may be modified into regulated structures or incorporate cascoding to improve performance [81, 82]. In order to achieve the best performance with a regulated common gate TIA, such as that proposed in [81], the DC output voltage of the TIA needs to close to the overdrive voltage, V_{DSAT}, of the input transistor and the load resistor needs to be approximately ten times larger than the drain to source resistance. Similar to CSPs, TIAs can also be made of technologies other than CMOS. The first TIA using organic thin-film transistors was proposed in [80], where the TIA was based on a voltage-controlled resistor and common gate input stage.

A multichannel readout based on a transimpedance topology was proposed in [74] (Fig. 7.6), where, in order to prevent the possible oscillations, a series resistor (R_s) was placed between the common anode of the SiPM and the input port of the operational amplifier. In this design, the inductors represent the parasitic inductances presented by the metal lines, wires, and printed circuit board. In order to modify the offset, output signal tail and the undershoot [83] uses trimmers. A TIA implemented in SiGe BiCMOS that used a transformer-based input stage for improved frequency response was proposed in response [84]. Figure 7.7 shows the gain and bandwidth of a few transimpedance amplifiers that have been implemented.

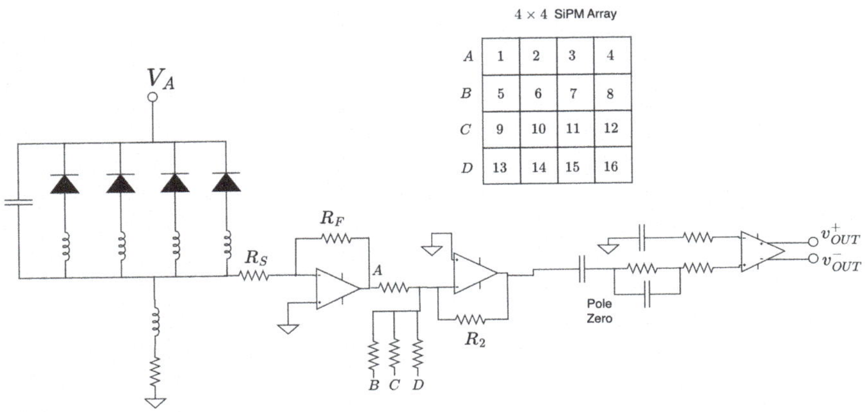

Fig. 7.6 A multichannel SiPM's readout based on transimpedance topology [74]

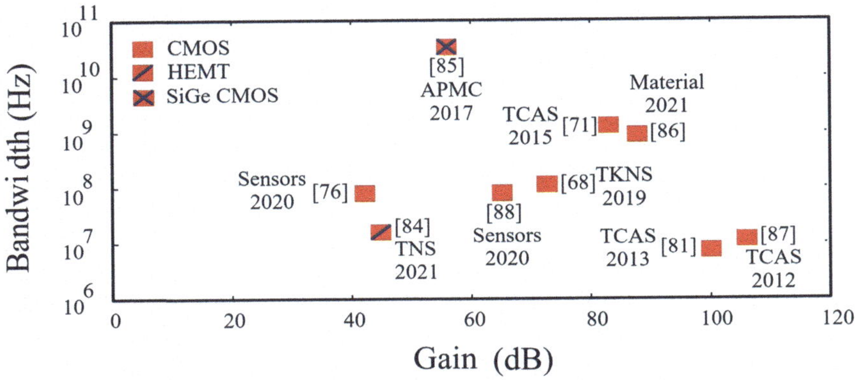

Fig. 7.7 Gain and bandwidth of transimpedance amplifiers proposed in [69, 72, 77, 82, 85–89]

7.2.3 *Voltage-Sensitive Preamplifiers*

Robustness, simplicity, and ease of implementation and simulation make the voltage amplifier an ideal choice of readout circuit when dealing with unknown SiPMs. However, this is sometimes achieved at the cost of implementing a large shunt resistor, compared to the TIA design. Large shunt resistors impose higher RC time constants which correspond with wider output pulses. In addition, the gain of this structure depends on the signal size and the currents flow through the device [74].

The use of a voltage amplifier requires a resistor in series with the SiPM. The voltage developed across the resistor is then amplified. A typical implementation using commercially available off-the-shelf parts may use the standard inverting or non-inverting configuration where the gain is a function of the resistive feedback network (Fig. 7.4c). The size of R_{load} affects the input DC bias of the

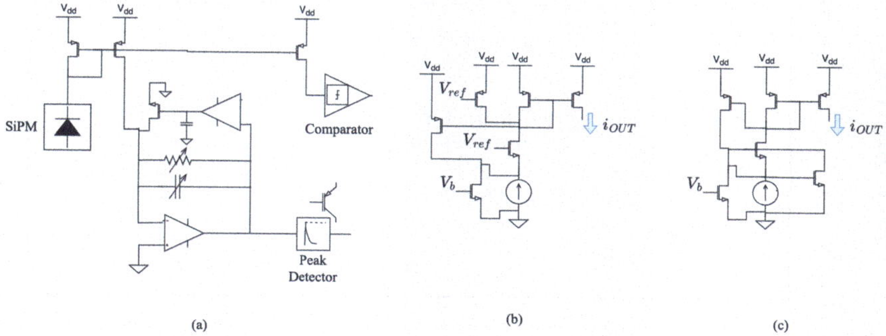

Fig. 7.8 (**a**) The current mode readout circuit proposed in [92]. Alternative current mode readout topologies for the current buffer (**b**, **c**) where i represents the SiPM current [93]

opamp [90]. The total input capacitance, formed by the parallel combination of parasitic capacitance of the SiPM and amplifier, C_d and C_{in}, and the total input impedance, forms a time constant that should be minimized such that the fastest input signal can be recognized [91].

In some applications further processing may require the output voltage to be converted back into current. To avoid multiple conversions, topologies such as that in Fig. 7.8a have been proposed [92]. To achieve better performance, the current mirror that is used in [92] can be replaced with improved topologies with higher output impedance by cascoding or regulated drain structures. These alternative topologies employing feedback such as Fig. 7.8b, c have also been implemented in 350-nm CMOS processes [93], where (b) achieves higher dynamic range and tunability and (c) is optimized for bandwidth and input impedance. In [78] the voltage amplifier is coupled with TIA. In this method, the width of the input current pulses is controlled by the circuit time constant which can be adjusted by TIA's feedback network. Figure 7.9 shows a comparison of the gain and bandwidth of voltage amplifiers implemented in the literature.

7.3 Energy Measurements

The photon energy of an event (defined as photons being incident on the SiPMs) is proportional to the SiPM's charge or current pulse and is an important characteristic. The peak signal provides information on the energy of the photons. However, regardless of readout type, energy resolution may be affected by the scintillator material, scintillator thickness [100], and other physical mechanisms such as ballistic deficit, pulse pileup, or baseline fluctuation [26, 101–105].

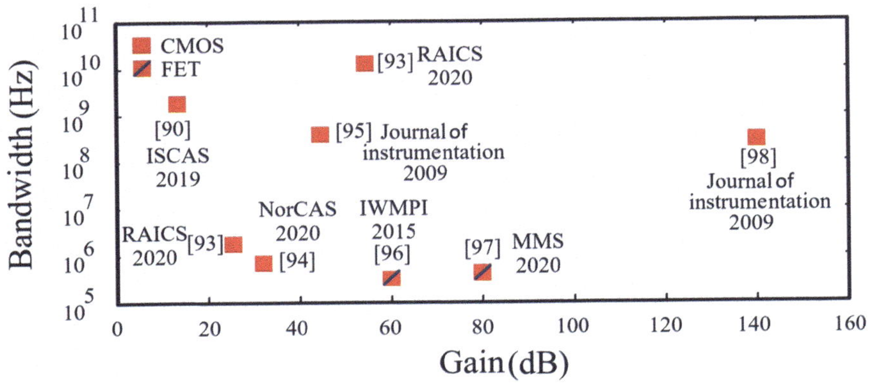

Fig. 7.9 Gain and bandwidth of the voltage amplifiers proposed in [91, 94–99]

Table 7.1 SiPM readout energy resolution based on scintillator type

Reference	[106] (2016)	[107] (2019)	[108] (2017)	[109] (2019)	[109] (2019)
Scintillator	LaBr3:Ce	CeBr3	LaBr3(Ce)	LaBr3(Ce)	LaBr3(Ce) co-doped with Sr
SiPM	Custom	SensL	FBK	FBK	FBK
Resolution	% 4 at 661 KeV	% 4.54 at 661 keV	% 3.19 at 661 keV	% 3.4 at 661 keV	% 2.6 at 661 keV

Table 7.1 shows a relative comparison of the energy resolution using different scintillator materials. The total photon absorption depends on the type and thickness of material and incident energy of photons. For example, four different scintillators, CsI(Tl), NaI(Tl), LaBr3(Ce), and GAGG(Ce), showed a direct relationship between energy resolution and material type and thickness [100]. Thus, it is important to take the scintillator material into account when characterizing the full system. In addition, since the scintillator is the front-end transducer, its properties will limit the overall imaging system characteristics. Figure 7.10 shows readout circuit topologies and scintillators.

Pulse pileup occurs when the time distance between the two or more pulses is less than the pulse resolution time [137]. Resolution time is the precision of an SiPM in determining the arrival time of a single photon. Pulse pileup can be either head pileup or tail pileup. Head pileup occurs when the rising edge of a second pulse comes to pass the rising edge of the first pulse. Tail pileup occurs when the rising edge of the second pulse occurs on the falling edge of the first pulse [138]. These pileups can result in a shift of the baseline voltage [55]. Pulse pileup can be mitigated by rejection, filtering, or a transformation algorithm. Many of these are implemented in software or using off-chip microprocessors and FPGAs.

The shaper stage (also known as shaper filters), which may be present in energy, timing, and position measurement [139, 140], is used to tackle the problem of pulse pileup. Gaussian-based architectures are popular shaper implementations [141].

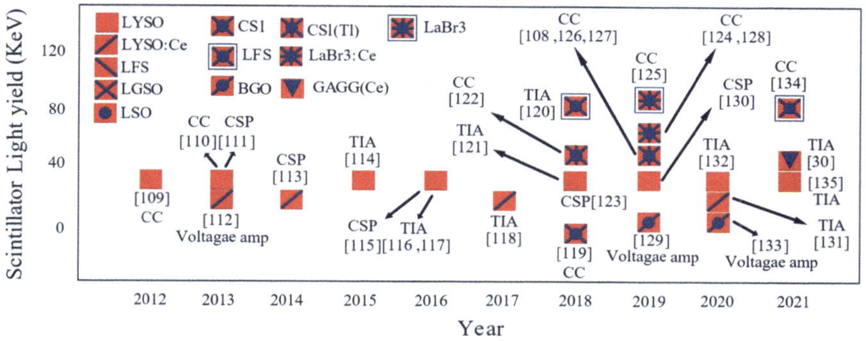

Fig. 7.10 Readout circuits used in nuclear imaging with various scintillator materials, from 2012 to 2021 [31, 109–136]

Frequency bandwidth limitations can improve the SNR [142]. However, bandwidth limitation may cause negative overshoot for large shaping parameter [139]. Quasi-Gaussian or trapezoidal shapers have also been implemented [140]. Low-pass filters, for example, using a Sallen–Key filter, can calibrate for drift and improve signal to noise [143]. Shapers have also been implemented using digital $CR - RC^m$ [144]. The pulse pileup problem needs to be investigated at high input rate application. This problem can be addressed by implementing pole-zero cancellation circuits (PZCs). An example of this type of circuit can be found in [55], where the PZC has been implemented by 20 PMOS transistors and a 24-pF capacitor. In addition, to tackle the problem of pileup, pole-zero cancellation eliminates the undershoot and allows for higher rate counting [140, 145–147].

The shaper's output signal is passed into a sample and hold circuit (also called a peak stretcher) to store the peak value of the signal before digitizing [148]. Analog to digital converters (ADCs) are also used in energy measurement systems. They range from successive approximation (SAR), delta-sigma, dual-slope, pipelined, and flash architectures [26, 149–151]. These architectures offer various advantages in speed, resolution, power, and area on chip. ADCs in general are power-hungry and can be area intensive on chip. Thus, other digitization and quasi-digitization strategies such as time-over-threshold (ToT) circuits may be employed [152]. In a ToT circuit, the difference between two time stamps where the signal crosses a predefined threshold is stored. This can be implemented with time to digital conversion [153, 154]. Integration of readout electronics and digital SiPM, within a single chip, can make them noisier than analog SiPMs. This is because the digital SiPMs will typically be in standard CMOS where no special efforts have been made to optimize the process for optical detection [155]. Many commercially available off-the-shelf SiPMs are analog and are optimized for optical response. For analog and digital SiPMs both implemented in CMOS, the noise of the optical detecting device remains the same and there are distinct advantages and disadvantages for the digital or analog approach. Customized analog SiPMs have better noise performance for energy measurements as the energy resolution depends on the optical performance, i.e., the statistical properties of photon-induced charge generation [156].

7.4 Timing Measurements

In addition to energy, the time of arrival of pulses, corresponding to events, is also of interest [157]. Figure 7.11 shows the response of various types of events. Time of flight measurements determine the distribution of the time difference between events. Uncorrelated signals will have a flat time difference distribution. Correlated signals, such as those which occur in prompt coincidence, delayed coincidence, or time of flight experiments, show a delta or spike-shaped distribution [158–160]. Noise in time of flight is controlled by the coincidence resolving time [161]. Time domain information can always be converted to frequency domain. This can be done either in software or in hardware using either analog, digital, or a combination of analog and digital techniques.

Figure 7.12a shows the use of a discriminator (comparator) to determine if an event occurred [162, 163]. Delay units to enable proper comparison of the time stamps [164, 165], time to amplitude converters (TACs) [166], and multichannel analyzers [167, 168] are required to accurately report the timing. Figure 7.12b filters the events within some range before the discriminator and improves the signal to noise [169]. Each of these approaches can be implemented using discrete off-the-shelf components or implemented using custom integrated circuits fabricated in commercial foundries. In all cases, the analog modules contain amplifiers and filters, along with the digital and quasi-digital blocks, such as comparators, and encoders [140, 170–174]. Depending on the scintillator type and material, such as LGSO, LYSO, GAGG, LuAG, YAG, LSO, and LFS, various types of

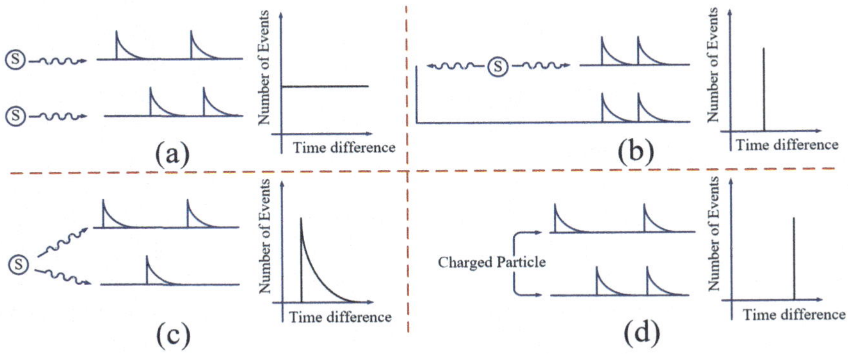

Fig. 7.11 Time measurements due to a source or sources, (**a**) uncorrelated events, (**b**) prompt coincidence, (**c**) delayed coincidence, and (**d**) time of flight [158]

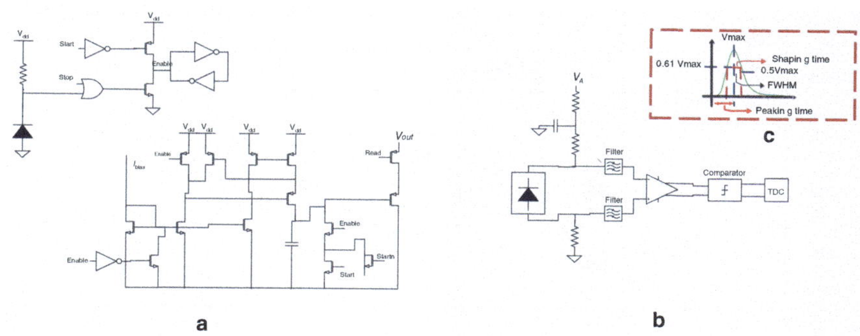

Fig. 7.12 (**a**) Time to analog conversion [180], (**b**) time spectrometer with a differential amplifier [169], and (**c**) definitions of the shaping and peaking times [181]

SiPMs may have different time resolutions [175]. The required resolution depends on application, where, for example, in PET, sub 10 ps is a research target. The maximum achievable coincidence time resolution (CTR), based on scintillator statistics, can be estimated using [176]:

$$CTR = \alpha\sqrt{\frac{\tau_r \tau_f}{n}} \tag{7.3}$$

where τ_r and τ_f are the scintillation rise and fall time, n is the number of photons, and α is a constant. For example, for an output light of between 27.9 and 49.5 kphoton/MeV measured upon gamma CS173 excitation, the best achievable CTR with cerium-doped GAGG:Ce scintillator was 87 $\pm$ 2 ps [162]. The other factor that has an influence on time resolution is the readout electronic noise. Cates et al. [157] proposed a passive compensation circuit for device capacitance to reduce the effect of electronic noise on time resolution. This technique provided no injection of noise from the SiPM to the readout circuit and improved the single-photon time resolution. In some other work, different time resolutions, ranging from less than 80 ps [175] up to more than 330 ps FWHM [177], can be found. In [178, 179], improved time resolution is achieved by connecting multiple SiPMs in series creating multiple scintillation counters. For a readout circuit, consisting of an N counter, the resolution improves by a factor of $1/\sqrt{N}$.

To determine the time of an event, comparators and digital gates may be used [182, 183]. Comparator type (or discriminators) can be constant fraction (CFD) or leading edge (LED) [184]. However at a lower threshold, LED provides better time resolution [185, 186]. Comparators can have errors due to jitter, long-term and short-term drift, and variation in the shape and amplitude of the input pulse [187]. This can be compensated for by 1-D techniques such as linear or logarithmic compensation or 2-D compensation techniques like artificial neural networks and polynomials [188, 189]. Time resolution is determined with the full width half

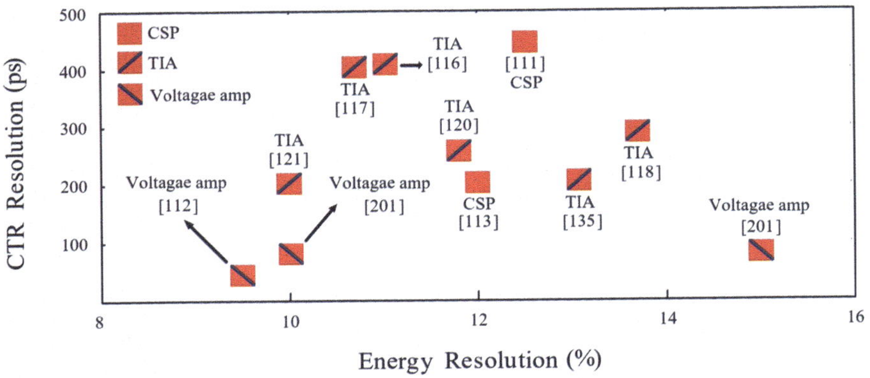

Fig. 7.13 Time and energy resolution of different readout topologies' charge amplifier (CSP), transimpedance amplifier (TIA), and voltage amplifier [112–114, 117–119, 121, 122, 136, 202]

maximum of the time spectrum [190]. To achieve the best time resolution [181], we suggest setting the shaping time at 61% of the pulse amplitude. where the peaking time needs to be approximately three to five times greater than the shaping time (Fig. 7.12c).

Delays can be generated using coaxial cables, active circuits, or lumped elements [149, 191, 192]. However, the use of coaxial cables may also attenuate, frequency shift, or reflect the signal. Time to amplitude converters (TACs) generate an analog output whose amplitude is proportional to the measured time interval. They may be implemented using a purely analog, digital, or mixed signal approach [180, 193, 194]. The time resolution is limited by the accuracy of the timing discriminators [146, 181]. As with the energy measurements, ADCs may be replaced by time to digital converters (TDCs) [195, 196].

Both analog and digital readout approaches can have a similar time resolution [197–200]. Due to increased power consumption, large numbers of readout channels, and challenges in digital realization [201], analog SiPMs are more prevalent in nuclear imaging applications. Figure 7.13 shows CTR and energy resolution for different front ends.

7.5 Position Sensing

When implemented in integrated circuits, the inherently asynchronous and event-based nature of the signal makes them suitable for employing the asynchronous readout such as in address event representation (AER) protocol [203–205]. This has been implemented in a standard CMOS technology with a digital SiPM architecture [206]. By comparison of the arrival time of two or more pixels, from the AER communication protocol, position information also can be extracted [207, 208] (Fig. 7.14).

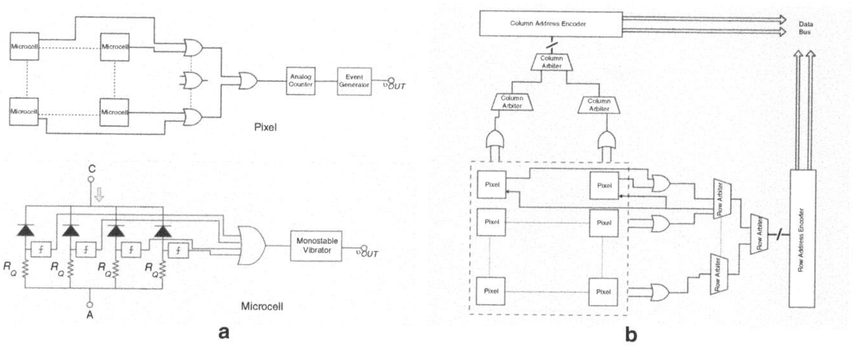

Fig. 7.14 (**a**) SPAD pixel within SiPM and (**b**) AER protocol [205]

For conventional analog SiPM arrays, typical position sensing may require a large number of readout channels and may show limited position resolution due to the SiPM's pixel size [209]. To achieve higher spatial resolution, artificial neural networks [132, 210] and novel position-sensitive SiPM (PS-SiPM) fabrication processes [211, 212] can increase the position sensing performance. Multiplexing [26, 213] and multichannel integrated readouts [165, 214, 215] simplify the readout and improve performance characteristics (Fig. 7.15). In this figure, each diode symbol represents an SiPM. Position detection has also been performed using multiple cameras, algorithms to determine the center of gravity of signals spread across the pixels, Compton scattering, and time of flight measurements [216–221]. In many cases these approaches involve significant use of signal processing software and thus approaches that utilize microprocessors on the same chip as the SiPM or with 3D integration can potentially offer improved readout speed and throughput.

7.6 Array Readout

For small arrays, each channel may be read independently. Each can have its own amplifier and data acquisition channel [222–225]. Larger arrays require multiplexing and processing to reduce complexity and cost [226, 227]. Multiplexing may reduce the capability to discriminate several events or scatter interactions and correct for anode/pixel nonuniformities [228]. Multiplexing of multiple SiPM sensors using resistor networks can lead to reduced timing performance [26]. Common multiplexing schemes include Anger, discretized positioning (DPC) [229], row/column or crosswire readout [230], and symmetric charge division multiplexing (SCD) [231]. Row/column readout can be standard X-Y or scrambled X-Y, where the devices are placed in series-parallel connections. In standard X-Y readout, all cathodes in each row and all anodes in each column are shorted. For a case with nine SiPM arrays, with each of those consisting of 16 pixels, the readout electronics

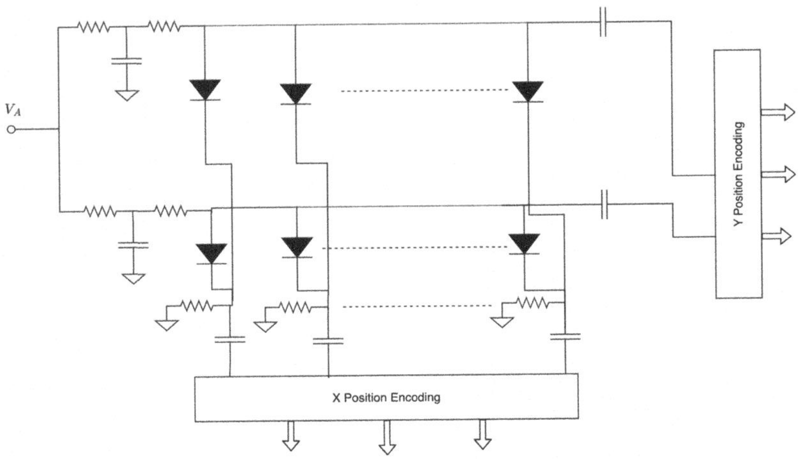

Fig. 7.15 Position sensing readout based on a multiplexing circuit [213]

require 12 threshold comparators to detect when an event has occurred. However, using a scrambled approach, where the cathodes of all the devices in each array are connected together, nine threshold detector channels can be used. Since the average energy, time, and position resolution is degraded with noise [213, 232], reduction in the number of SiPMs connected to a single readout channel results in less noise and better system performance [233].

7.7 Conclusions

The readout architecture of an SiPM-based array will be a function of application. Applications range from those which have weak light sources to those with stronger incident light sources. Depending on the application, either digital or analog SiPMs can be used. Digital SiPMs digitize at the microcell level, after which any digital system such as an FPGA, complex logic, or microprocessor can be used to process the signal. For many applications, an analog SiPM is used instead due to the increased optical active area. The readout of analog SiPMs typically targets energy, time, or position of interaction. The readout electronics can be implemented using commercial off-the-shelf components or custom integrated circuits. Both options add constraints that must be taken into account for a given application and SPICE models may be used to predict system performance. Discrete electronics will typically have larger parasitics and thus operate at slower speeds than systems where the devices and electronics are integrated on the same chip.

The primary challenges in SiPM readout design can be broken down into three categories, the readout circuit, packaging and interfaces, and signal processing.

In general, an imaging system consists of hundreds of SiPMs with many readout circuits that are typically multiplexed to reduce the power and complexity. However, multiplexing can impair the SiPM's performance. Strategies may involve sharing circuits between pixels or adding in circuits for calibration. At the packaging and interface level misalignment and edge effects may affect energy, time, and position resolution. Post fabrication of a scintillator material directly on top of the SiPM is a possible solution to this. However, wafer dicing is subject to inaccuracies and new materials and fabrication protocols are germane to any integration of a scintillator and optical detector. Processing of the signal can be performed after digitizing using analog to digital converters, microprocessors, and software. These are all very power-hungry and space-intensive. Readout using in-pixel digitized approaches along with asynchronous readout can help reduce complexity. Finally, the use of 3D integration can allow for increased pixel fill factor (i.e., increased optical active area) while still allowing integrated circuit-sized readout electronics.

References

1. S.H. Hesari, M.A. Haque, N. McFarlane, A comprehensive survey of readout strategies for SiPMs used in nuclear imaging Systems. Photonics **8**, 266 (2021). https://doi.org/10.3390/photonics8070266
2. C. Alispach, J. Borkowski, F.R. Cadoux, N. De Angelis, D. Della Volpe, Y. Favre, M. Heller, J. Juryšek, E. Lyard, D. Mandat et al., Large scale characterization and calibration strategy of a SiPM-based camera for gamma-ray astronomy. J. Instrum. **15**, P11010 (2020)
3. F. Henningsen, M. Boehmer, A. Gärtner, L. Geilen, R. Gernhäuser, H. Heggen, K. Holzapfel, C. Fruck, L. Papp, I. Rea et al., A self-monitoring precision calibration light source for large-volume neutrino telescopes. J. Instrum. **15**, P07031 (2020)
4. Y. Sun, J. Maricic, SiPMs characterization and selection for the DUNE far detector photon detection system. J. Instrum. **11**, C01078 (2016)
5. G. Adamo, A. Busacca, Time Of Flight measurements via two LiDAR systems with SiPM and APD, in *Proceedings of the AEIT International Annual Conference*, Capri, 5–7 Oct 2016, pp. 1–5
6. S. Gnecchi, S. Buckley, S. Bellis, C. Barry, C. Jackson, SiPM sensors for TOF ranging applications, in *Proceedings of the SPIE Photonics West*, San Francisco, 13–18 Feb 2016
7. S. Gnecchi, C. Barry, S. Bellis, S. Buckley, C. Jackson, Long distance ranging performance of Gen3 LiDAR imaging system based on 1 × 16 SiPM array, in *Proceedings of the International Image Sensor Workshop*, Snowbird, 23–27 June 2019
8. S. Gnecchi, C. Jackson, A 1 × 16 SiPM array for automotive 3D imaging LiDAR systems, in *Proceedings of the International Image Sensor Workshop*, Hiroshima, 30 May–2 June 2017, pp. 133–136
9. M. Jeong, M. Hammig, Development of hand-held coded-aperture gamma ray imaging system based on GAGG (Ce) scintillator coupled with SiPM array. Nucl. Eng. Technol. **52**, 2572–2580 (2020)
10. K. Shimazoe, M. Yoshino, Y. Ohshima, M. Uenomachi, K. Oogane, T. Orita, H. Takahashi, K. Kamada, A. Yoshikawa, M. Takahashi, Development of simultaneous PET and Compton imaging using GAGG-SiPM based pixel detectors. Nucl. Instrum. Methods Phys. Res. Sect. A Accel. Spectrom. Detect. Assoc. Equip. **954**, 161499 (2020)
11. Z. Liang, T. Hu, X. Li, Y. Wu, C. Li, Z. Tang, A cosmic ray imaging system based on plastic scintillator detector with SiPM readout. J. Instrum. **15**, C07033 (2020)

12. G. Maira, A.M. Chiarelli, S. Brafa, S. Libertino, G. Fallica, A. Merla, S. Lombardo, Imaging system based on silicon photomultipliers and light emitting diodes for functional near-infrared spectroscopy. Appl. Sci. **10**, 1068 (2020)
13. V.J. Kitsmiller, T.D. O'Sullivan, Next-generation frequency domain diffuse optical imaging systems using silicon photomultipliers. Opt. Lett. **44**, 562–565 (2019)
14. S. Dey, E. Myers, T.K. Lewellen, R.S. Miyaoka, J.C. Rudell, A row-column summing readout architecture for SiPM based PET imaging systems, in *Proceedings of the IEEE Nuclear Science Symposium and Medical Imaging Conference*, Seoul, 27 Oct–2 Nov 2013, pp. 1–5
15. J.W. Cates, M.F. Bieniosek, C.S. Levin, Highly multiplexed signal readout for a time-of-flight positron emission tomography detector based on silicon photomultipliers. J. Med. Imaging **4**, 011012 (2017)
16. H. Kim, C.M. Kao, Y. Hua, Q. Xie, C.T. Chen, Multiplexing readout for time-of-flight (TOF) PET detectors using striplines. IEEE Trans. Radiat. Plasma Med. Sci. (2021). https://doi.org/10.1109/TRPMS.2021.3051364
17. Q. Yang, Z. Kuang, Z. Sang, Y. Yang, J. Du Performance comparison of two signal multiplexing readouts for SiPM-based PET detector. Phys. Med. Biol. **64**, 23NT02 (2019)
18. M.A. Tétrault, E.D. Lamy, A. Boisvert, C. Thibaudeau, M. Kanoun, F. Dubois, R. Fontaine, J.F. Pratte, Real-time discrete SPAD array readout architecture for time of flight PET. IEEE Trans. Nucl. Sci. **62**, 1077–1082 (2015)
19. M.G. Giacomelli, Evaluation of silicon photomultipliers for multiphoton and laser scanning microscopy. J. Biomed. Opt. **24**, 106503 (2019)
20. C. Martin, T. Li, E. Hegarty, P. Zhao, S. Mondal, A. Ben-Yakar, Line excitation array detection fluorescence microscopy at 0.8 million frames per second. Nat. Commun. **9**, 1–10 (2018)
21. P. Dorosz, M. Baszczyk, W. Kucewicz, The study of single cells in a system based on Silicon Photomultipliers, in *Proceedings of the IEEE Nuclear Science Symposium and Medical Imaging Conference*, Sydney, 10–17 Nov 2018, pp. 1–2
22. M. Heller, E. Schioppa Jr., A. Porcelli, I.T. Pujadas, K. Zietara, D. Della Volpe, T. Montaruli, F. Cadoux, Y. Favre, J. Aguilar et al., An innovative silicon photomultiplier digitizing camera for gamma-ray astronomy. Eur. Phys. J. C **77**, 1–31 (2017)
23. J. Zhang, R. Zhou, S. Zhang, H. Xiong, G. Hu, Y. Li, Z. Liu, C. Yang, Data acquisition and trigger system for imaging atmospheric Cherenkov telescopes of the LHAASO. J. Instrum. **15**, T02004 (2020)
24. H. Xiong, R. Zhou, J. Chen, Y. Li, C. Yang, B. Bi, J. Zhang, J. Zhang, X. Cao, S. Zhang et al., Design and performance of analog circuit for the wide field of view Cherenkov telescope array of LHAASO. Nucl. Instrum. Methods Phys. Res. Sect. A Accel. Spectrom. Detect. Assoc. Equip. **925**, 156–163 (2019)
25. A. Berti, A. Chiavassa, D. Corti, D. Depaoli, F. Di Pierro, L. Lessio, M. Mallamaci, M. Mariotti, C. Perennes, R. Rando et al., Development and test of a SiPM cluster for a SiPM version of the Cherenkov Telescope Array LST camera. Nucl. Instrum. Methods Phys. Res. Sect. A Accel. Spectrom. Detect. Assoc. Equip. **982**, 164373 (2020)
26. S. Jiang, J. Lu, S. Meng, F. Ye, J. Ning, Q. Yi, J. Yang, R. Yang, X. Yan, A prototype of SiPM-based scintillator Compton camera with capacitive multiplexing readout. J. Instrum. **16**, P01027 (2021)
27. S. Aldawood, P. Thirolf, A. Miani, M. Böhmer, G. Dedes, R. Gernhäuser, C. Lang, S. Liprandi, L. Maier, T. Marinšek et al., Development of a Compton camera for prompt-gamma medical imaging. Radiat. Phys. Chem. **140**, 190–197 (2017)
28. Y. Nakamura, K. Shimazoe, H. Takahashi, S. Yoshimura, Y. Seto, S. Kato, M. Takahashi, T. Momose, Development of a novel handheld intra-operative laparoscopic Compton camera for 18F-Fluoro-2-deoxy-2-D-glucose-guided surgery. Phys. Med. Biol. **61**, 5837 (2016)
29. H. Lee, T. Lee, W. Lee, Development of a position-sensitive 4π Compton camera based on a single segmented scintillator. IEEE Trans. Nucl. Sci. **67**, 2511–2522 (2020)
30. G. Cozzi, P. Busca, M. Carminati, C. Fiorini, A. Gola, C. Piemonte, V. Regazzoni, Development of a SiPM-based detection module for prompt gamma imaging in proton therapy, in *Proceedings of the IEEE Nuclear Science Symposium, Medical Imaging Conference and*

Room-Temperature Semiconductor Detector Workshop, Strasbourg, 29 Oct–6 Nov 2016, pp. 1–5
31. J. Boo, M.D. Hammig, M. Jeong, Row–column readout method to mitigate radiographic-image blurring from multipixel events in a coded-aperture imaging system. IEEE Trans. Nucl. Sci. **68**, 1175–1183 (2021)
32. E. Martinenghi, L. Di Sieno, D. Contini, M. Sanzaro, A. Pifferi, A. Dalla Mora, Time-resolved single-photon detection module based on silicon photomultiplier: a novel building block for time-correlated measurement systems. Rev. Sci. Instrum. **87**, 073101 (2016)
33. B. Li, Q. Miao, S. Wang, D. Hui, T. Zhao, K. Liang, R. Yang, D. Han, Time-correlated photon counting (TCPC) technique based on a photon-number-resolving photodetector. Adv. Photon Count. Tech. X **9858**, 98580L (2016)
34. G. Acconcia, A. Cominelli, M. Ghioni, I. Rech, Fast fully-integrated front-end circuit to overcome pile-up limits in time-correlated single photon counting with single photon avalanche diodes. Opt. Express **26**, 15398–15410 (2018)
35. F. Corsi, A. Dragone, C. Marzocca, A. Del Guerra, P. Delizia, N. Dinu, C. Piemonte, M. Boscardin, G.F. Dalla Betta, Modelling a silicon photomultiplier (SiPM) as a signal source for optimum front-end design. Nucl. Instrum. Methods Phys. Res. Sect. A Accel. Spectrom. Detect. Assoc. Equip. **572**, 416–418 (2007)
36. D. Marano, M. Belluso, G. Bonanno, S. Billotta, A. Grillo, S. Garozzo, G. Romeo, O. Catalano, G. La Rosa, G. Sottile et al., Silicon photomultipliers electrical model extensive analytical analysis. IEEE Trans. Nucl. Sci. **61**, 23–34 (2013)
37. F. Licciulli, C. Marzocca, Parameter Extraction Method for the Electrical Model of a Silicon Photomultiplier. IEEE Trans. Nucl. Sci. **63**, 2517–2526 (2016)
38. J. Du, X. Bai, A. Gola, F. Acerbi, A. Ferri, C. Piemonte, Y. Yang, S.R. Cherry, Performance of a high-resolution depth-encoding PET detector module using linearly-graded SiPM arrays. Phys. Med. Biol. **63**, 035035 (2018)
39. A. Muntean, A. Sachdeva, E. Venialgo, S. Gnecchi, D. Palubiak, C. Jackson, E. Charbon, A fully integrated state-of-the-art analog SiPM with on-chip time conversion, in *Proceedings of the IEEE Nuclear Science Symposium and Medical Imaging Conference*, Sydney, 10–17 Nov 2018, pp. 1–3
40. R.J. Walker, E.A. Webster, J. Li, N. Massari, R.K. Henderson, High fill factor digital silicon photomultiplier structures in 130nm CMOS imaging technology, in *Proceedings of the IEEE Nuclear Science Symposium and Medical Imaging Conference Record*, Anaheim, 27 Oct–3 Nov 2012, pp. 1945–1948
41. G. Paternoster, L. Ferrario, F. Acerbi, A.G. Gola, P. Bellutti, Silicon photomultipliers technology at Fondazione Bruno Kessler and 3D integration perspectives, in *Proceedings of the European Solid-State Device Research Conference*, Cracow, 23–26 Sep 2019, pp. 50–53
42. E. Vilella, O. Alonso, A. Dieguez, 3D integration of Geiger-mode avalanche photodiodes aimed to very high fill-factor pixels for future linear colliders. Nucl. Instrum. Methods Phys. Res. Sect. A Accel. Spectrom. Detect. Assoc. Equip. **731**, 103–108 (2013)
43. G. Romanchek, Y. Wang, H. Marupudi, S. Abbaszadeh, Performance of optical coupling materials in scintillation detectors post temperature exposure. Sensors **20**, 6092 (2020)
44. M. Saukoski, L. Aaltonen, K. Halonen, T. Salo, Fully integrated charge sensitive amplifier for readout of micromechanical capacitive sensors, in *Proceedings of the IEEE International Symposium on Circuits and Systems*, Kobe, 23–26 May 2005, pp. 5377–5380
45. Y. Song, J. Rozsa, J. Magalhaes, S. Smith, B. Karlinsey, W. Kinnison, E. Gustafson, D. Austin, A.R. Hawkins, S.H.W. Chiang, Design of common-mode feedback for high-gain charge amplifier, in *Proceedings of the Intermountain Engineering, Technology and Computing Conference*, Orem, 2–3 Oct 2020, pp. 1–5
46. P. Kmon, G. Deptuch, F. Fahim, P. Grybos, P. Maj, R. Szczygiel, T. Zimmerman, Active feedback with leakage current compensation for charge sensitive amplifier used in hybrid pixel detector. IEEE Trans. Nucl. Sci. **66**, 664–673 (2019)
47. R. Bassini, C. Boiano, A. Pullia, A low-noise charge amplifier with fast rise time and active discharge mechanism. IEEE Trans. Nucl. Sci. **49**, 2436–2439 (2002)

48. M. Manghisoni, D. Comotti, L. Gaioni, L. Ratti, V. Re, Dynamic compression of the signal in a charge sensitive amplifier: experimental results. IEEE Trans. Nucl. Sci. **65**, 636–644 (2017)
49. E. Alnasser, A novel low output offset voltage charge amplifier for piezoelectric sensors. IEEE Sensors J. **20**, 5360–5367 (2020)
50. S. Jeong, D. Shin, J. Kim, A transformer-isolated common-mode active EMI filter using a low-cost BJT amplifier with feedforward structure, in *Proceedings of the International Conference on Power Electronics and ECCE Asia*, Busan, 27–30 May 2019, pp. 2088–2094
51. D.J. Comer, D.T. Comer, Bipolar junction transistor amplifiers, in *Fundamentals of Circuits and Filters* (CRC Press, Boca Raton, 2018)
52. P. Durdaut, V. Penner, C. Kirchhof, E. Quandt, R. Knochel, M. Hoft, Noise of a JFET charge amplifier for piezoelectric sensors. IEEE Sensors J. **17**, 7364–7371 (2017)
53. G. Scandurra, G. Giusi, C. Ciofi, Single JFET front-end amplifier for low frequency noise measurements with cross correlation-based gain calibration. Electronics **8**, 1197 (2019)
54. Z. He, C. Wang, G. Fan, Y. Zhou, Y. Yang, Design of a high input impedance OPA with Bi-JFET technology, in *Proceedings of the IEEE International Conference on Electronics Technology*, Chengdu, 10–13 May 2019, pp. 233–236
55. P. Grybos, M. Idzik, K. Swientek, P. Maj, Integrated charge sensitive amplifier with pole-zero cancellation circuit for high rates, in *Proceedings of the IEEE International Symposium on Circuits and Systems*, Kos, 21–24 May 2006, p. 4
56. K. Kasinski, R. Kleczek, A flexible, low-noise charge-sensitive amplifier for particle tracking application, in *Proceedings of the 2016 MIXDES-23rd International Conference Mixed Design of Integrated Circuits and Systems*, Lodz, 23–25 June 2016, pp. 124–129
57. M. Manghisoni, D. Comotti, L. Gaioni, L. Ratti, V. Re, Dynamic compression of the signal in a charge sensitive amplifier: from concept to design. IEEE Trans. Nucl. Sci. **62**, 2318–2326 (2015)
58. C. Lee, G. Cho, T. Unruh, S. Hur, I. Kwon, Integrated circuit design for radiation-hardened charge-sensitive amplifier survived up to 2 Mrad. Sensors **20**, 2765 (2020)
59. M. Manghisoni, L. Ratti, V. Re, V. Speziali, G. Traversi, G. Fallica, Gamma-ray response of SOI bipolar junction transistors for fast, radiation tolerant front-end electronics. Nucl. Instrum. Methods Phys. Res. Sect. A Accel. Spectrom. Detect. Assoc. Equip. **518**, 477–481 (2004)
60. T.D. England, Silicon-Germanium BiCMOS and Silicon-on-Insulator CMOS Analog Circuits for Extreme Environment Applications. Ph.D. Thesis, Georgia Institute of Technology, Atlanta, 2014
61. R.R. Harrison, C. Charles, A low-power low-noise CMOS amplifier for neural recording applications. IEEE J. Solid-State Circuits **38**, 958–965 (2003)
62. P. Giannelli, G. Calabrese, G. Frattini, M. Granato, L. Capineri, A buffered single-supply charge amplifier for high-impedance piezoelectric sensors. IEEE Trans. Instrum. Meas. **68**, 368–376 (2018)
63. S. Capra, A. Pullia, Design and experimental validation of an integrated multichannel charge amplifier for solid-state detectors with innovative spectroscopic range booster. IEEE Trans. Nucl. Sci. **67**, 1877–1884 (2020)
64. I. Yadav, H. Shrimali, Noise and crosstalk models of the particle detector with zero-pole transformation charge sensitive amplifier. Nucl. Instrum. Methods Phys. Res. Sect. A Accel. Spectrom. Detect. Assoc. Equip. **937**, 107–116 (2019)
65. I. Yadav, A. Joshi, E. Ruscino, V. Liberali, A. Andreazza, H. Shrimali, Design of a charge sensitive amplifier for silicon particle detector in BCD 180 nm process, in *Proceedings of the 2019 32nd International Conference on VLSI Design and 2019 18th International Conference on Embedded Systems*, Delhi, 5–9 Jan 2019, pp. 541–542
66. M. Beikahmadi, S. Mirabbasi, A low-power low-noise CMOS charge-sensitive amplifier for capacitive detectors, in *Proceedings of the IEEE International New Circuits and Systems Conference*, Bordeaux, 26–29 June 2011, pp. 450–453
67. Z. Gu, X. Bi, A charge amplifier with noise peaking suppression and gain drop compensation utilizing a Quasi-Miller RC network. AEU-Int. J. Electron. Commun. **107**, 252–256 (2019)

68. I. Kwon, T. Kang, B.T. Wells, J. Lawrence, M.D. Hammig, Compensation of the detector capacitance presented to charge-sensitive preamplifiers using the Miller effect. Nucl. Instrum. Methods Phys. Res. Sect. A Accel. Spectrom. Detect. Assoc. Equip. **784**, 220–225 (2015)
69. M. Seung, W. Choi, S. Hur, I. Kwon, Design of a transimpedance amplifier for a radiation detector to observe a nuclear reactor under severe accidents, in *Proceedings of the Transactions of the Korean Nuclear Society Spring Meeting*, Jeju, 23–24 May 2019
70. F. Zocca, A. Pullia, G. Pascovici, Design and optimization of low-noise wide-bandwidth charge preamplifiers for high purity germanium detectors. IEEE Trans. Nucl. Sci. **55**, 695–702 (2008)
71. H.M. Ip, S. Thomas, M. Hart, M. Prydderch, M. French, A dynamic slew correction circuit for low noise silicon detector pre-amplifiers. IEEE Trans. Nucl. Sci. **59**, 642–646 (2012)
72. I. Kwon, T. Kang, B.T. Wells, J. Lawrence, M.D. Hammig, A high-gain 1.75-GHz dual-inductor transimpedance amplifier with gate noise suppression for fast radiation detection. IEEE Trans. Circuits Syst. II Express Briefs **63**, 356–360 (2015)
73. M.V. Montenegro, Front-End CMOS Transimpedance Amplifiers on a Silicon Photomultiplier Resistant to Fast Neutron Fluence. Ph.D. Thesis, University of Nevada, Las Vegas, 2020
74. J. Kaspar, A. Fienberg, D. Hertzog, M. Huehn, P. Kammel, K. Khaw, D. Peterson, M. Smith, T. Van Wechel, A. Chapelain et al., Design and performance of SiPM-based readout of PbF2 crystals for high-rate, precision timing applications. J. Instrum. **12**, P01009 (2017)
75. G. Barile, A. Leoni, L. Pantoli, L. Safari, V. Stornelli, A new VCII based low-power low-voltage front-end for silicon photomultipliers, in *Proceedings of the International Conference on Smart and Sustainable Technologies*, Split, 26–29 June 2018, pp. 1–4
76. L. Pantoli, G. Barile, A. Leoni, M. Muttillo, V. Stornelli, A novel electronic interface for micromachined Si-based photomultipliers. Micromachines **9**, 507 (2018)
77. V. Stornelli, L. Pantoli, G. Barile, A. Leoni, E. D'Amico, Silicon photomultiplier sensor interface based on a discrete second generation voltage conveyor. Sensors **20**, 2042 (2020)
78. M. Ren, J. Zhou, B. Song, C. Zhang, M. Dong, R. Albarracín, Towards optical partial discharge detection with micro silicon photomultipliers. Sensors **17**, 2595 (2017)
79. T.L. Naing, T.O. Rocheleau, E. Alon, C.T.C. Nguyen, A 78-microwatt GSM phase noise-compliant pierce oscillator referenced to a 61-MHz wine-glass disk resonator, in *Proceedings of the Joint European Frequency and Time Forum & International Frequency Control Symposium*, Prague, 21–25 July 2013, pp. 562–565
80. S. Elsaegh, H. Zappe, Y. Manoli, H. Klauk, U. Zschieschang, A 1.6 μW tunable organic transimpedance amplifier for photodetector applications based on gain-boosted common-gate input stage and voltage-controlled resistor with $\pm 0.5\%$ nonlinearity, in *Proceedings of the European Solid State Circuits Conference*, Leuven, 11–14 Sep 2017, pp. 75–78
81. D.C. Dias, F.S. de Melo, L.B. Oliveira, J.P. Oliveira, Regulated common-gate transimpedance amplifier for radiation detectors and receivers, in *Proceedings of the International Conference Mixed Design of Integrated Circuits and Systems*, Lublin, 19–21 June 2014, pp. 540–545
82. M. de Medeiros Silva, L.B. Oliveira, Regulated common-gate transimpedance amplifier designed to operate with a silicon photo-multiplier at the input. IEEE Trans. Circuits Syst. I Regul. Pap. **61**, 725–735 (2013)
83. G. Ambrosi, M. Ambrosio, C. Aramo, E. Bissaldi, A. Boiano, C. Bonavolontà, C. de Lisio, L. Di Venere, E. Fiandrini, N. Giglietto et al., Development of a charge preamplifier to improve NUV-HD SiPM performances. Nucl. Part. Phys. Proc. **291**, 40–43 (2017)
84. A. Karimi-Bidhendi, H. Mohammadnezhad, M.M. Green, P. Heydari, A silicon-based low-power broadband transimpedance amplifier. IEEE Trans. Circuits Syst. I Regul. Pap. **65**, 498–509 (2017)
85. V.K. Pandey, C.M. Tan, Application of gallium nitride technology in particle therapy imaging. IEEE Trans. Nucl. Sci. (2021). https://doi.org/10.1109/TNS.2021.3072654
86. S. Giannakopoulos, Z.S. He, I. Darwazeh, H. Zirath, Differential common base TIA with 56 dB Ohm gain and 45 GHz bandwidth in 130 nm SiGe, in *Proceedings of the IEEE Asia Pacific Microwave Conference*, Kuala Lumpur, 13–16 Nov 2017, pp. 1107–1110

87. S. Kodam, A. Arashanapalli, J. Kolla, G. Dheeraj, Design and implementation of CMOS transimpedance amplifier as a photodetector. Mater. Today Proc. (2021). https://doi.org/10.1016/j.matpr.2021.01.204
88. L.B. Oliveira, C.M. Leitão, M.M. Silva, Noise performance of a regulated cascode transimpedance amplifier for radiation detectors. IEEE Trans. Circuits Syst. I Regul. Pap. **59**, 1841–1848 (2012)
89. S.H. Hesari, A. Hedayatipour, M.A. Haque, N. McFarlane, A bulk driven transimpedance amplifier for portable SiPM based detectors, in *Proceedings of the 2020 IEEE Sensors*, Rotterdam, 25–28 Oct 2020, pp. 1–4
90. J. Huizenga, S. Seifert, F. Schreuder, H. Van Dam, P. Dendooven, H. Löhner, R. Vinke, D. Schaart, A fast preamplifier concept for SiPM-based time-of-flight PET detectors. Nucl. Instrum. Methods Phys. Res. Sect. A Accel. Spectrom. Detect. Assoc. Equip. **695**, 379–384 (2012)
91. M. Wang, Y. Wang, Q. Cao, L. Wang, J. Kuang, Y. Xiao, Comparison of three pre-amplifier circuits for time readout of SiPM in TOF-PET detectors, in *Proceedings of the International Symposium on Circuits and Systems*, Sapporo, 26–29 May 2019, pp. 1–5
92. A. Argentieri, F. Corsi, M. Foresta, C. Marzocca, A. Del Guerra, Design and characterization of CMOS multichannel front-end electronics for silicon photomultipliers. Nucl. Instrum. Methods Phys. Res. Sect. A Accel. Spectrom. Detect. Assoc. Equip. **652**, 516–519 (2011)
93. F. Corsi, M. Foresta, C. Marzocca, G. Matarrese, A. Del Guerra, Current-mode front-end electronics for silicon photo-multiplier detectors, in *Proceedings of the International Workshop on Advances in Sensors and Interface*, Bari, 26–27 June 2007, pp. 1–6
94. V. Lakshmy, E. Shajahan, Performance analysis of low power low voltage amplifier designs in 45 nm CMOS technology by incorporating Miller compensation, in *Proceedings of the 2020 IEEE Recent Advances in Intelligent Computational Systems (RAICS)*, Thiruvananthapuram, 3–5 Dec 2020, pp. 18–22
95. L.H. Rodovalho, Schmitt trigger based single-ended voltage amplifier with positive feedback control for ultra-low-voltage supplies, in *Proceedings of the 2020 IEEE Nordic Circuits and Systems Conference (NorCAS)*, Oslo, 27–28 Oct 2020, pp. 1–6
96. N. Dinu, P. Barrillon, C. Bazin, N. Belcari, M.G. Bisogni, S. Bondil-Blin, M. Boscardin, V. Chaumat, G. Collazuol, C. De La Taille et al., Characteristics of a prototype matrix of Silicon PhotoMultipliers (SiPM). J. Instrum. **4**, P03016 (2009)
97. W. Zhang, B. Zheng, P. Goodwill, S. Conolly, A custom low-noise preamplifier for Magnetic Particle Imaging, in *Proceedings of the 2015 5th International Workshop on Magnetic Particle Imaging (IWMPI)*, Istanbul, 26–28 Mar 2015, p. 1
98. K. Achtenberg, J. Mikołajczyk, Z. Bielecki, FET input voltage amplifier for low frequency noise measurements. Metrol. Meas. Syst. 531–540 (2020). https://doi.org/10.24425/mms.2020.132785
99. F. Corsi, M. Foresta, C. Marzocca, G. Matarrese, A. Del Guerra, ASIC development for SiPM readout. J. Instrum. **4**, P03004 (2009)
100. J.M. Brown, In-silico optimisation of tileable Philips digital SiPM based thin monolithic scintillator detectors for SPECT applications. Appl. Radiat. Isot. **168**, 109368 (2021)
101. B. Kim, K. Lee, J. Lee, S. Hwang, D. Heo, K. Han, Design of optimal digital filter and digital signal processing for a CdZnTe high resolution gamma-ray system. Appl. Radiat. Isot. **162**, 109171 (2020)
102. M. Sammartini, M. Gandola, F. Mele, B. Garavelli, D. Macera, P. Pozzi, G. Bertuccio, X-ray spectroscopy with a CdTe pixel detector and SIRIO preamplifier at deep submicrosecond signal-processing time. IEEE Trans. Nucl. Sci. **68**, 70–75 (2020)
103. M. Lee, D. Lee, E. Ko, K. Park, J. Kim, K. Ko, M. Sharma, G. Cho, Pulse pileup correction method for gamma-ray spectroscopy in high radiation fields. Nucl. Eng. Technol. **52**, 1029–1035 (2020)
104. P. Adrich, I. Zychor, Least-squares fitting algorithm for peak pile-up correction in gamma-ray spectroscopy, in *Nuclear Instruments and Methods in Physics Research Section A: Accelerators, Spectrometers, Detectors and Associated Equipment* (Elsevier, Amsterdam, 2020), p. 164962

105. M. Pocaterra, M. Ciappa, Experimental setup to monitor non-destructive single events triggered by ionizing radiation in power devices. Microelectron. Reliab. **114**, 113755 (2020)
106. A. Ulyanov, O. Morris, L. Hanlon, S. McBreen, S. Foley, O.J. Roberts, I. Tobin, D. Murphy, C. Wade, N. Nelms et al., Performance of a monolithic LaBr3: Ce crystal coupled to an array of silicon photomultipliers. Nucl. Instrum. Methods Phys. Res. Sect. A Accel. Spectrom. Detect. Assoc. Equip. **810**, 107–119 (2016)
107. C. Lavelle, W. Shanks, C. Chiang, M. Nichols, J. Osborne Jr., A. Herschelman, B. Brown, M. Cho, Approaches for single channel large area silicon photomultiplier array readout. AIP Adv. **9**, 035123 (2019)
108. G. Cozzi, P. Busca, M. Carminati, C. Fiorini, G.L. Montagnani, F. Acerbi, A. Gola, G. Paternoster, C. Piemonte, V. Regazzoni et al., High-resolution gamma-ray spectroscopy with a SiPM-based detection module for 1" and 2" LaBr 3: Ce readout. IEEE Trans. Nucl. Sci. **65**, 645–655 (2017)
109. G. Montagnani, L. Buonanno, D. Di Vita, M. Carminati, F. Camera, A. Gola, V. Regazzoni, C. Fiorini, Spectroscopic performance of a Sr co-doped 3" LaBr3 scintillator read by a SiPM array. Nucl. Instrum. Methods Phys. Res. Sect. A Accel. Spectrom. Detect. Assoc. Equip. **931**, 158–161 (2019)
110. M.D. Rolo, R. Bugalho, F. Gonçalves, A. Rivetti, G. Mazza, J.C. Silva, R. Silva, J. Varela, A 64-channel ASIC for TOFPET applications, in *Proceedings of the IEEE Nuclear Science Symposium and Medical Imaging Conference Record*, Anaheim, 27 Oct–3 Nov 2012, pp. 1460–1464
111. M. Rolo, R. Bugalho, F. Goncalves, G. Mazza, A. Rivetti, J. Silva, R. Silva, J. Varela, Tofpet asic for pet applications. J. Instrum. **8**, C02050 (2013)
112. I. Sacco, P. Fischer, M. Ritzert, PETA4: a multi-channel TDC/ADC ASIC for SiPM readout. J. Instrum. **8**, C12013 (2013)
113. J. Fleury, S. Callier, C. de La Taille, N. Seguin, D. Thienpont, F. Dulucq, S. Ahmad, G. Martin, Petiroc, a new front-end ASIC for time of flight application, in *Proceedings of the IEEE Nuclear Science Symposium and Medical Imaging Conference*, Seoul, 27 Oct–2 Nov 2013, pp. 1–5
114. H. Chen, K. Briggl, P. Fischer, A. Gil, T. Harion, Y. Munwes, M. Ritzert, D. Schimansky, H.C. Schultz-Coulon, W. Shen et al., A dedicated readout ASIC for time-of-flight positron emission tomography using silicon photomultiplier (SiPM), in *Proceedings of the IEEE Nuclear Science Symposium and Medical Imaging Conference*, Seattle, 8–15 Nov 2014, pp. 1–5
115. S. Ahmad, J. Fleury, C. de la Taille, N. Seguin-Moreau, F. Dulucq, G. Martin-Chassard, S. Callier, D. Thienpont, L. Raux, Triroc: a multi-channel SiPM read-out ASIC for PET/PET-ToF application. IEEE Trans. Nucl. Sci. **62**, 664–668 (2015)
116. R. Becker, C. Casella, S. Corrodi, G. Dissertori, J. Fischer, A. Howard, M. Ito, W. Lustermann, Studies of the high rate coincidence timing response of the STiC and TOFPET ASICs for the SAFIR PET scanner. J. Instrum. **11**, P12001 (2016)
117. S. Ahmad, C. de la Taille, J. Fleury, N. Seguin-Moreau, L. Raux, S. Callier, G. Martin-Chassard, F. Dulucq, D. Thienpont, Triroc, a versatile 64-channel SiPM readout ASIC for time-of-flight PET, in *Proceedings of the IEEE Nuclear Science Symposium, Medical Imaging Conference and Room-Temperature Semiconductor Detector Workshop*, Strasbourg, 29 Oct–6 Nov 2016, pp. 1–5
118. G. Sportelli, S. Ahmad, N. Belcari, M.G. Bisogni, N. Camarlinghi, A. Di Pasquale, S. Dussoni, J. Fleury, M. Morrocchi, E. Zaccaro et al., The TRIMAGE PET data acquisition system: initial results. IEEE Trans. Radiat. Plasma Med Sci. **1**, 168–177 (2016)
119. N. Omidvari, J. Cabello, G. Topping, F.R. Schneider, S. Paul, M. Schwaiger, S.I. Ziegler, PET performance evaluation of MADPET4: a small animal PET insert for a 7 T MRI scanner. Phys. Med. Biol. **62**, 8671 (2017)
120. P. Trigilio, P. Busca, R. Quaglia, M. Occhipinti, C. Fiorini, A SiPM-readout ASIC for SPECT applications. IEEE Trans. Radiat. Plasma Med Sci. **2**, 404–410 (2018)
121. A.L. Goertzen, D. Van Elburg, Performance characterization of MPPC modules for TOF-PET applications. IEEE Trans. Radiat. Plasma Med Sci. **3**, 475–482 (2018)

122. D. Schug, V. Nadig, B. Weissler, P. Gebhardt, V. Schulz, Initial measurements with the PETsys TOFPET2 ASIC evaluation kit and a characterization of the ASIC TDC. IEEE Trans. Radiat. Plasma Med Sci. **3**, 444–453 (2018)
123. M. Carminati, G.L. Montagnani, M. Occhipinti, A. Kuehne, T. Niendorf, K. Nagy, A. Nagy, M. Czeller, C. Fiorini, SPECT/MRI INSERT compatibility: assessment, solutions, and design guidelines. IEEE Trans. Radiat. Plasma Med Sci. **2**, 369–379 (2018)
124. Z. Zhao, S. Xie, X. Zhang, J. Yang, Q. Huang, J. Xu, Q. Peng, An advanced 100-channel readout system for nuclear imaging. IEEE Trans. Instrum. Meas. **68**, 3200–3210 (2018)
125. M. Carminati, F. Baratelli, M. Occhipinti, K. Erlandsson, K. Nagy, Z. Nyitrai, M. Czeller, A. Kühne, T. Niendorf, S. Valtorta et al., Validation and performance assessment of a preclinical SiPM-based SPECT/MRI insert. IEEE Trans. Radiat. Plasma Med Sci. **3**, 483–490 (2019)
126. L. Buonanno, G. Montagnani, D. Di Vita, F. Castelli, M. Carminati, C. Fiorini, GAMMA: a high dynamic range 16-ch ASIC for large scintillators readout with SiPM array, in *Proceedings of the IEEE Nuclear Science Symposium and Medical Imaging Conference*, Manchester, 26 Oct–2 Nov 2019, pp. 1–3
127. D. Di Vita, L. Buonanno, G. Montagnani, A. Minerva, A. Giannoni, M. Carminati, F. Camera, C. Fiorini, A high dynamic range 144-SiPM detection module for gamma spectroscopy and imaging with 3" LaBr 3, in *Proceedings of the IEEE Nuclear Science Symposium and Medical Imaging Conference*, Manchester, 26 Oct–2 Nov 2019, pp. 1–3
128. G.L. Montagnani, L. Buonanno, D. Di Vita, C. Fiorini, M. Carminati, A compact 4-decade dynamic range readout module for gamma spectroscopy and imaging, in *Proceedings of the 2019 IEEE International Symposium on Circuits and Systems (ISCAS)*, Sapporo, 26–29 May 2019, pp. 1–5
129. M. Carminati, I. D'Adda, A. Morahan, K. Erlandsson, K. Nagy, M. Czeller, B. Tölgyesi, Z. Nyitrai, A. Savi, P. van Mullekom et al., Clinical SiPM-based MRI-compatible SPECT: preliminary characterization. IEEE Trans. Radiat. Plasma Med Sci. **4**, 371–377 (2019)
130. X. Yue, C. Liu, Z. Deng, An energy-resolved photon-counting readout electronics for scintillator based on pole-zero compensation and ToT method, in *Proceedings of the IEEE Nuclear Science Symposium and Medical Imaging Conference*, Manchester, 26 Oct–2 Nov 2019, pp. 1–3
131. X. Cheng, K. Hu, Y. Shao, Dual-polarity SiPM readout electronics based on 1-bit sigma-delta modulation circuit for PET detector applications. IEEE Trans. Nucl. Sci. **66**, 2107–2113 (2019)
132. A. LaBella, P. Vaska, W. Zhao, A.H. Goldan, Convolutional neural network for crystal identification and gamma ray localization in PET. IEEE Trans. Radiat. Plasma Med. Sci. **4**, 461–469 (2020)
133. J. Barrio, N. Cucarella, A.J. Gonzalez, M. Freire, V. Ilisie, J.M. Benlloch, Characterization of a high-aspect ratio detector with lateral sides readout for Compton PET. IEEE Trans. Radiat. Plasma Med. Sci. **4**, 546–554 (2020)
134. N. Kratochwil, E. Auffray S. Gundacker, Exploring Cherenkov emission of BGO for TOF-PET. IEEE Trans. Radiat. Plasma Med. Sci. **5**(5), 619–629 (2021). https://doi.org/10.1109/TRPMS.2020.3030483
135. D. Sánchez, S. Gómez, J.M. Fernández-Tenllado, R. Ballabriga, M. Campbell, D. Gascón, Multimodal simulation of large area silicon photomultipliers for time resolution optimization. Nucl. Instrum. Methods Phys. Res. Sect. A Accel. Spectrom. Detect. Assoc. Equip. **1001**, 165247 (2021)
136. V. Nadig, D. Schug, B. Weissler, V. Schulz, Evaluation of the PETsys TOFPET2 ASIC in multi-channel coincidence experiments. EJNMMI Phys. **8**, 1–21 (2021)
137. G. Contino, O. Catalano, G. Sottile, P. Sangiorgi, M. Capalbi, G. Osteria, V. Scotti, H. Miyamoto, C. Vigorito, M. Casolino et al., An ASIC front-end for fluorescence and Cherenkov light detection with SiPM for space and ground applications. Nucl. Instrum. Methods Phys. Res. Sect. A Accel. Spectrom. Detect. Assoc. Equip. **980**, 164510 (2020)
138. S. Molodtsov, A. Gurbich, Simulation of the pulse pile-up effect on the pulse-height spectrum. Nucl. Instrum. Methods Phys. Res. Sect. B Beam Interact. Mater. Atoms **267**, 3484–3487 (2009)

139. H. Wu, H. Xu, Q. Zhang, X. Meng, R. Di, Y. Liu, Implementation of a straight-line shaper on an FPGA for a UAVAGS. J. Instrum. **15**, T10011 (2020)
140. Y. Liu, J. Zhang, L. Liu, S. Li, R. Zhou, Implementation of real-time digital CR–RCm shaping filter on FPGA for gamma-ray spectroscopy. Nucl. Instrum. Methods Phys. Res. Sect. A Accel. Spectrom. Detect. Assoc. Equip. **906**, 1–9 (2018)
141. M.Y. Kantor, A. Sidorov, Shaping pulses of radiation detectors into a true Gaussian form. J. Instrum. **14**, P01004 (2019)
142. G. Zeng, J. Yang, T. Hu, L. Ge, X. Ouyang, Q. Zhang, Y. Gu, Baseline restoration technique based on symmetrical zero-area trapezoidal pulse shaper. Nucl. Instrum. Methods Phys. Res. Sect. A Accel. Spectrom. Detect. Assoc. Equip. **858**, 57–61 (2017)
143. S. Saxena, A.I. Hawari, Investigation of FPGA-based real-time adaptive digital pulse shaping for high-count-rate applications. IEEE Trans. Nucl. Sci. **64**, 1733–1738 (2017)
144. J. Wang, X. Luo, X. Wei, R. Zheng, C. Feng, Y. Hu, Design of a low-noise, high-linearity, readout ASIC for CdZnTe-based gamma-ray spectrometers. Nucl. Instrum. Methods Phys. Res. Sect. A Accel. Spectrom. Detect. Assoc. Equip. **988**, 164927 (2021)
145. C. Zhao, Z. Wang, M. Zhao, C. Yang, Characteristics and comparison of two digital quasi-Gauss filters for gamma spectroscopy. J. Instrum. **16**, P01021 (2021)
146. S. Saxena, A.I. Hawari, Digital pulse deconvolution with adaptive shaping for real-time high-resolution high-throughput gamma spectroscopy. Nucl. Instrum. Methods Phys. Res. Sect. A Accel. Spectrom. Detect. Assoc. Equip. **954**, 161288 (2020)
147. C. Liu, X. Luo, R. Zheng, J. Wang, X. Wei, F. Xue, Y. Hu, A low noise APD readout ASIC for electromagnetic calorimeter in HIEPA. Nucl. Instrum. Methods Phys. Res. Sect. A Accel. Spectrom. Detect. Assoc. Equip. **985**, 164686 (2021)
148. M. Carminati, D. Di Vita, L. Buonanno, G.L. Montagnani, C. Fiorini, A lightweight SiPM-based gamma-ray spectrometer for environmental monitoring with drones, in *International Conference on Applications in Electronics Pervading Industry, Environment and Society* (Springer, Cham, 2020), pp. 55–61
149. S. Krishnan, C. Webster, A. Duffy, G. Brooks, R. Clay, J. Mould, Improving the energy resolution while mitigating the effects of dark-noise, for a microcontroller based SiPM sensor. J. Instrum. **15**, P09028 (2020)
150. Y. Tang, R. Zhang, J. Chen, A low-power 16-channel SiPM readout front-end with a shared SAR ADC in 180 nm CMOS, in *Proceedings of the International Conference on Solid-State & Integrated Circuit Technology*, Kunming, 3–6 Nov 2020, pp. 1–3
151. H. Zhang, H. Xu, Y. Fan, X. Xing, H. Liu, J. Wu, A time-interleaved pipelined ADC with ultra high speed sampling. Sādhanā **45**, 1–8 (2020)
152. H. Wang, J. Liu, H. Zhang, H. Liang, B. Ye, Multiple time-over-threshold readout electronics for fast timing and energy resolving in a SiPM-based positron annihilation lifetime spectrometer. J. Instrum. **15**, P11006 (2020)
153. M.D. Zhu, X. Qin, L. Wang, W.Z. Zhang, Y. Lin, X. Rong, J. Du, A time-to-digital-converter utilizing bits-counters to decode carry-chains and DSP48E1 slices in a field-programmable-gate-array. J. Instrum. **16**, P02009 (2021)
154. J.F. Pratte, F. Nolet, S. Parent, F. Vachon, N. Roy, T. Rossignol, K. Deslandes, H. Dautet, R. Fontaine, S.A. Charlebois, 3D photon-to-digital converter for radiation instrumentation: motivation and future works. Sensors **21**, 598 (2021)
155. S. Gundacker, A. Heering, The silicon photomultiplier: fundamentals and applications of a modern solid-state photon detector. Phys. Med. Biol. **65**, 17TR01 (2020)
156. M.N. Ullah, J.H. Park, E. Pratiwi, G.B. Kim, J.Y. Yeom, Wavelength discrimination (WLD) detector optimization for time-of-flight positron emission tomography with depth of interaction information. Nucl. Instrum. Methods Phys. Res. Sect. A Accel. Spectrom. Detect. Assoc. Equip. **982**, 164498 (2020)
157. J.W. Cates, S. Gundacker, E. Auffray, P. Lecoq, C.S. Levin, Improved single photon time resolution for analog SiPMs with front end readout that reduces influence of electronic noise. Phys. Med. Biol. **63**, 185022 (2018)
158. M. Nakhostin, *Signal Processing for Radiation Detectors* (Wiley, Hoboken, 2017)

159. P. Lecoq, S. Gundacker, SiPM applications in positron emission tomography: toward ultimate PET time-of-flight resolution. Eur. Phys. J. Plus **136**, 1–16 (2021)
160. S. Surti, J.S. Karp, Update on latest advances in time-of-flight PET. Phys. Med. **80**, 251–258 (2020)
161. H. Wieczorek, A. Thon, T. Dey, V. Khanin, P. Rodnyi, Analytical model of coincidence resolving time in TOF-PET. Phys. Med. Biol. **61**, 4699 (2016)
162. L. Martinazzoli, N. Kratochwil, S. Gundacker, E. Auffray, Scintillation properties and timing performance of state-of-the-art Gd3Al2Ga3O12 single crystals. Nucl. Instrum. Methods Phys. Res. Sect. A Accel. Spectrom Detect. Assoc. Equip. **1000**, 165231 (2021)
163. A. Rivetti, M. Alexeev, R. Bugalho, F. Cossio, M.D.R. Rolo, A. Di Francesco, M. Greco, W. Cheng, M. Maggiora, S. Marcello et al., TIGER: a front-end ASIC for timing and energy measurements with radiation detectors. Nucl. Instrum. Methods Phys. Res. Sect. A Accel. Spectrom. Detect. Assoc. Equip. **924**, 181–186 (2019)
164. D. Impiombato, A. Segreto, O. Catalano, S. Giarrusso, T. Mineo, Use of the Peak-Detector mode for gain calibration of SiPM sensors with ASIC CITIROC read-out. J. Instrum. **15**, C04007 (2020)
165. M. Parsakordasiabi, I. Vornicu, Á. Rodríguez-Vázquez, R. Carmona-Galán, A Low-Resources TDC for Multi-Channel Direct ToF Readout Based on a 28-nm FPGA. Sensors **21**, 308 (2021)
166. A. Lobanov, Precision timing calorimetry with the CMS HGCAL. J. Instrum. **15**, C07003 (2020)
167. A.T. Susanto, P. Prajitno, S.K. Wijaya, S. Soekirno, I.P. Susila, A systematic literature reviews of multichannel analyzer based on FPGA for gamma spectroscopy. J. Phys. Conf. Ser. **1528**, 012016 (2020)
168. J.A. Schmitz, M.K. Gharzai, S. Balkir, M.W. Hoffman, M. Bauer, A low-power 10-bit multichannel analyzer chip for radiation detection, in *Proceedings of the International Symposium on Circuits and Systems*, Baltimore, 28–31 May 2017, pp. 1–4
169. F. Powolny, E. Auffray, S. Brunner, E. Garutti, M. Goettlich, H. Hillemanns, P. Jarron, P. Lecoq, T. Meyer, H. Schultz-Coulon et al., Time-based readout of a silicon photomultiplier (SiPM) for time of flight positron emission tomography (TOF-PET). IEEE Trans. Nucl. Sci. **58**, 597–604 (2011)
170. Z.J. Qin, C. Chen, J.S. Luo, X.H. Xie, L.Q. Ge, Q.F. Wu, A pulse-shape discrimination method for improving Gamma-ray spectrometry based on a new digital shaping filter. Radiat. Phys. Chem. **145**, 193–201 (2018)
171. C. Sundberg, M.U. Persson, M. Sjölin, J.J. Wikner, M. Danielsson, Silicon photon-counting detector for full-field CT using an ASIC with adjustable shaping time. J. Med Imaging **7**, 053503 (2020)
172. R.R. Jogdand, P. Dakhole, P. Palsodkar, Low power flash ADC using multiplexer based encoder, in *Proceedings of the International Conference on Innovations in Information, Embedded and Communication Systems*, Coimbatore, 17–18 Mar 2017, pp. 1–5
173. V.K. Pandey, C.M. Tan, S. Sharma, Design of GaN based comparator circuit for radiation detectors, in *Proceedings of the India Council International Conference*, New Delhi, 10–13 Dec 2020, pp. 1–6
174. L. Qiu, K. Wang, K. Tang, L. Siek, Y. Zheng, A 10-bit 300 MS/s 5.8 mW SAR ADC with two-stage interpolation for PET imaging. IEEE Sensors J. **18**, 2006–2014 (2018)
175. S. Gundacker, F. Acerbi, E. Auffray, A. Ferri, A. Gola, M. Nemallapudi, G. Paternoster, C. Piemonte, P. Lecoq, State of the art timing in TOF-PET detectors with LuAG, GAGG and L (Y) SO scintillators of various sizes coupled to FBK-SiPMs. J. Instrum. **11**, P08008 (2016)
176. S. Gundacker, R. Turtos, E. Auffray, P. Lecoq, Precise rise and decay time measurements of inorganic scintillators by means of X-ray and 511 keV excitation. Nucl. Instrum. Methods Phys. Res. Sect. A Accel. Spectrom. Detect. Assoc. Equip. **891**, 42–52 (2018)
177. M. Nemallapudi, S. Gundacker, P. Lecoq, E. Auffray, Single photon time resolution of state of the art SiPMs. J. Instrum. **11**, P10016 (2016)

178. J. Zhang, M.I. Knopp, M.V. Knopp, Sparse detector configuration in SiPM digital photon counting PET: a feasibility study. Mol. Imaging Biol. **21**, 447–453 (2019)
179. P. Cattaneo, M. De Gerone, F. Gatti, M. Nishimura, W. Ootani, M. Rossella, S. Shirabe, Y. Uchiyama, Time resolution of time-of-flight detector based on multiple scintillation counters readout by SiPMs. Nucl. Instrum. Methods Phys. Res. Sect. A Accel. Spectrom. Detect. Assoc. Equip. **828**, 191–200 (2016)
180. Z. Wu, Y. Xu, Z. Ma, A time-to-amplitude converter with high impedance switch topology for single-photon time-of-flight measurement. IEEE Access **9**, 16672–16678 (2021)
181. E. Ermis, C. Celiktas, Time resolution investigations for general purpose plastic scintillation detectors in different thicknesses. J. Radioanal. Nucl. Chem. **295**, 523–536 (2013)
182. G. Akamatsu, S. Takyu, E. Yoshida, Y. Iwao, H. Tashima, F. Nishikido, T. Yamaya, Evaluation of a Hamamatsu TOF-PET detector module with 3.2 mm pitch LFS scintillators and a 256-channel SiPM array. IEEE Trans. Radiat. Plasma Med Sci. (2020). https://doi.org/10.1109/TRPMS.2020.3035876
183. M. Sadik, X. Ai, Y. Lu, R. Nock, Real-time time correlated photon counters for photon number resolving detectors. Opt. Sens. Detect. VI **11354**, 1135406 (2020)
184. D.R. Schaart, Physics and technology of time-of-flight PET detectors. Phys. Med. Biol. **66**(9), 09TR01 (2021)
185. C.M. Du, J.D. Chen, X.L. Zhang, H.B. Yang, K. Cheng, J. Kong, Z.G. Hu, Z.Y. Sun, H. Su, H.S. Xu, Study of time resolution by digital methods with a DRS4 module. Chin. Phys. C **40**, 046101 (2016)
186. T. Xu, J. Wen, Q. Wang, Q. Wei, T. Ma, Y. Liu, Y. Tai, A novel sub-millimeter resolution PET detector with TOF capability, in *Proceedings of the Nuclear Science Symposium and Medical Imaging Conference*, Seoul, 27 Oct–2 Nov 2013, pp. 1–5
187. H. Zhang, X. Wang, L. Zhang, Z. Zhang, A 10-bit 120-MS/s SAR ADC in 90 nm CMOS with redundancy compensation, in *Proceedings of the MTT-S International Wireless Symposium*, Chengdu, 6–10 May 2018, pp. 1–3
188. V. Bespal'ko, I. Burak, K. Salmins, Estimating the precision of a leading-edge discriminator with amplitude correction. Instrum. Exp. Tech. **62**, 788–793 (2019)
189. S. Xie, X. Zhang, Q. Huang, Z. Gong, J. Xu, Q. Peng, Methods to compensate the time walk errors in timing measurements for PET detectors. IEEE Trans. Radiat. Plasma Med Sci. **4**, 555–562 (2020)
190. A. Oberstedt, A. Gatera, A. Göök, S. Oberstedt, Time response and its impact on prompt fission γ-ray spectra characteristics. Eur. Phys. J. A **56**, 1–8 (2020)
191. H. Wang, Q. Zhao, H. Liang, B. Gu, J. Liu, H. Zhang, B. Ye, A new SiPM-based positron annihilation lifetime spectrometer using LYSO and LFS-3 scintillators. Nucl. Instrum. Methods Phys. Res. Sect. A Accel. Spectrom. Detect. Assoc. Equip. **960**, 163662 (2020)
192. P. Calò, S. Petrignani, C. Marzocca, B. Markovic, A. Dragone, A CMOS front-end for timing and charge readout of silicon photomultipliers, in *Proceedings of the Nuclear Science Symposium and Medical Imaging Conference*, Manchester, 26 Oct–2 Nov 2019, pp. 1–5
193. J. Yang, Y. Li, Y. Tian, Y. Fu, L. Xu, Y. Cai, Y. Li, Performance optimization of pixelated CdZnTe detector readout by analog ASIC using cathode waveform. J. Instrum. **15**, P04005 (2020)
194. Q. Yang, X. Wang, Z. Kuang, C. Zhang, Y. Yang, J. Du, Evaluation of two SiPM arrays for depth-encoding PET detectors based on dual-ended readout. IEEE Trans. Radiat. Plasma Med. Sci. (2020). https://doi.org/10.1109/TRPMS.2020.3008710
195. N.G. Abdallah, M. Rashdan, A.A. Khalaf, High resolution time-to-digital converter for PET imaging, in *Proceedings of the International Conference on Innovative Trends in Communication and Computer Engineering*, Aswan, 8–9 Feb 2020, pp. 295–298
196. H. Chen, D.D.U. Li, Multichannel, low nonlinearity time-to-digital converters based on 20 and 28 nm FPGAs. IEEE Trans. Ind. Electron. **66**, 3265–3274 (2018)
197. S. Gundacker, E. Auffray, P. Jarron, T. Meyer, P. Lecoq, On the comparison of analog and digital SiPM readout in terms of expected timing performance. Nucl. Instrum. Methods Phys. Res. Sect. A Accel. Spectrom. Detect. Assoc. Equip. **787**, 6–11 (2015)

198. A. Brogna, F. D'Andria, C. Marzocca, X. Selmani, Q. Weitzel, Time-performance design and study of ultra-wideband amplifiers for SiPM. J. Instrum. **15**, C02015 (2020)
199. P. Lecoq, C. Morel, J.O. Prior, D. Visvikis, S. Gundacker, E. Auffray, KrižP. an, R.M. Turtos, D. Thers, E. Charbon et al., Roadmap toward the 10 ps time-of-flight PET challenge. Phys. Med. Biol. **65**, 21RM01 (2020)
200. S.M. Decker, M. Pizzichemi, A. Polesel, M. Paganoni, E. Auffray, S. Gundacker, The digital-analog SiPM approach: a story of electronic and excess noise, in *Proceedings of the IEEE Nuclear Science Symposium and Medical Imaging Conference*, Manchester, 26 Oct–2 Nov 2019, pp. 1–5
201. H.G. Kang, T. Yamaya, Y.B. Han, S.H. Song, G.B. Ko, J.S. Lee, S.J. Hong, Crystal surface and reflector optimization for the SiPM-based dual-ended readout TOF-DOI PET detector. Biomed. Phys. Eng. Express **6**, 065028 (2020)
202. M.S. Lee, J.W. Cates, M. Kapusta, M. Schmand, C.S. Levin, Study of Lutetium-based scintillators for PET system design with 100-ps coincidence time resolution, in *Proceedings of the IEEE Nuclear Science Symposium and Medical Imaging Conference*, Manchester, 26 Oct–2 Nov 2019, pp. 1–2
203. A. Berkovich, T. Datta-Chaudhuri, P. Abshire, A scalable 20 × 20 fully asynchronous SPAD-based imaging sensor with AER readout, in *Proceedings of the International Symposium on Circuits and Systems*, Lisbon, 24–27 May 2015, pp. 1110–1113
204. J. Lin, A.G. Andreou, A 32 × 32 single photon avalanche diode imager with delay-insensitive address-event readout, in *Proceedings of the International Symposium of Circuits and Systems*, Rio de Janeiro, 15–18 May 2011, pp. 1824–1827
205. M.S.A. Shawkat, N. McFarlane, A CMOS Perimeter Gated SPAD Based Digital Silicon Photomultiplier with Asynchronous AER Readout for PET Applications, in *Proceedings of the IEEE Biomedical Circuits and Systems Conference*, Cleveland, 17–19 Oct 2018, pp. 1–4
206. M.S.A. Shawkat, N. Mcfarlane, A digital CMOS silicon photomultiplier using perimeter gated single photon avalanche diodes with asynchronous AER readout. IEEE Trans. Circuits Syst. I Regul. Pap. **67**, 4818–4828 (2020)
207. J.A. Leñero-Bardallo, R. Carmona-Galán, A. Rodríguez-Vázquez, Applications of event-based image sensors—Review and analysis. Int. J. Circuit Theory Appl. **46**, 1620–1630 (2018)
208. J.A. Lenero-Bardallo, F. Pérez-Pena, R. Carmona-Galán, Á. Rodríguez-Vázquez, Pipeline AER arbitration with event aging, in *Proceedings of the International Symposium on Circuits and Systems*, Baltimore, 28–31 May 2017, pp. 1–4
209. M. Jeong, B. Van, B.T. Wells, L.J. D'Aries, M.D. Hammig, Scalable gamma-ray camera for wide-area search based on silicon photomultipliers array. Rev. Sci. Instrum. **89**, 033106 (2018)
210. A. Taheri, M. Askari, M. Sasanpour, A beta-gamma position-sensitive detector based on rod plastic scintillators and artificial neural networks. J. Instrum. **15**, P06008 (2020)
211. Y. Peng, W. Lv, L. Dai, T. Zhao, K. Liang, R. Yang, D. Han, A square-bordered position-sensitive silicon photomultiplier toward distortion-free performance with high spatial resolution. IEEE Electron Device Lett. **41**, 1802–1805 (2020)
212. S. Kumar, M. Herzkamp, D. Durini, H. Nöldgen, S. van Waasen, Development of a solid-state position sensitive neutron detector prototype based on 6Li-glass scintillator and digital SiPM arrays. Nucl. Instrum. Methods Phys. Res. Sect. A Accel. Spectrom. Detect. Assoc. Equip. **954**, 161697 (2020)
213. J. Du, X. Bai, S.R. Cherry, A depth-encoding PET detector for high resolution PET using 1 mm SiPMs. Phys. Med. Biol. **65**, 165011 (2020)
214. Y. Chen, Z. Deng, Y. Liu, DIET: a multi-channel SiPM readout ASIC for TOF-PET with individual energy and timing digitizer. J. Instrum. **13**, P07023 (2018)
215. E.M. Zannoni, M.D. Wilson, K. Bolz, M. Goede, F. Lauba, D. Schöne, J. Zhang, M.C. Veale, M. Verhoeven, L.J. Meng, Development of a multi-detector readout circuitry for ultrahigh energy resolution single-photon imaging applications. Nucl. Instrum. Methods Phys. Res. Sect. A Accel. Spectrom. Detect. Assoc. Equip. **981**, 164531 (2020)

216. M. Ljungberg, P.H. Pretorius, SPECT/CT: an update on technological developments and clinical applications. Br. J. Radiol. **91**, 20160402 (2018)
217. D. Turecek, J. Jakubek, E. Trojanova, L. Sefc, Single layer Compton camera based on Timepix3 technology. J. Instrum. **15**, C01014 (2020)
218. A.A. Attarwala, D. Hardiansyah, C. Romanó, L.D. Jiménez-Franco, M. Roscher, B. Wängler, G. Glatting, Performance assessment of the ALBIRA II pre-clinical SPECT S102 system for 99 m Tc imaging. Ann. Nucl. Med. **35**, 111–120 (2021)
219. X.L. Sun, H. Wang, X.K. Li, G.H. Cao, Y. Kuang, X.C. Zhang, Monte Carlo computer simulation of a camera system for proton beam range verification in cancer treatment. Future Gener. Comput. Syst. **102**, 978–991 (2020)
220. C. Desmonts, M.A. Bouthiba, B. Enilorac, C. Nganoa, D. Agostini, N. Aide, Evaluation of a new multipurpose whole-body CzT-based camera: comparison with a dual-head Anger camera and first clinical images. EJNMMI Phys. **7**, 1–16 (2020)
221. K. Fujieda, J. Kataoka, S. Mochizuki, L. Tagawa, S. Sato, R. Tanaka, K. Matsunaga, T. Kamiya, T. Watabe, H. Kato et al., First demonstration of portable Compton camera to visualize 223-Ra concentration for radionuclide therapy. Nucl. Instrum. Methods Phys. Res. Sect. A Accel. Spectrom. Detect. Assoc. Equip. **958**, 162802 (2020)
222. WB Series Silicon Photomultipliers (SiPM)—KETEK GmbH, Available online: https://www.ketek.net/sipm/sipm-modules/tia-modules/ (Accessed on 2 Apr 2021)
223. PETsys SiPM Readout System, Available online: https://www.petsyselectronics.com/web/public/products/6 (Accessed on 4 Feb 2021)
224. A1702—32-Channel SiPM Readout Front-End Board—CAEN—Tools for Discovery, Available online: https://www.caen.it/products/a1702/ (Accessed on 4 Feb 2021)
225. PhotoniQ Multi-Channel Data Acquisition Systems, Available online: https://vertilon.com/pdf/UM6177.pdf (Accessed on 4 Feb 2021)
226. A. Selfridge, S. Cherry, R. Badawi, Characterization of four readout circuits for an MR compatible, preclinical PET detector. Phys. Med. Biol. **65**, 125008 (2020)
227. C. Mihai, G. Pascovici, G. Ciocan, C. Costache, V. Karayonchev, A. Lungu, N. Mărginean, R. Mihai, C. Neacşu, J.M. Régis et al., Development of large area Silicon Photomultipliers arrays for γ-ray spectroscopy applications. Nucl. Instrum. Methods Phys. Res. Sect. A Accel. Spectrom. Detect. Assoc. Equip. **953**, 163263 (2020)
228. H. Poladyan, O. Bubon, A. Teymurazyan, S. Senchurov, A. Reznik, Gaussian position-weighted center of gravity algorithm for multiplexed readout. Phys. Med. Biol. **65**, 165003 (2020)
229. S.J. Jeon, J.S. Lee, S.H. Hong, D.H. Kim, Y.W. Choi, Analysis of position error generated in readout resistive network for gamma-ray detection system, in *Proceedings of the International Conference on Electronics, Information, and Communication*, Barcelona, 19–22 Jan 2020, pp. 1–3
230. J. Boo, M. Hammig, M. Jeong, Hand-Held Dual-Particle Imager Implemented with a Multiplexed Low Sampling-Rate Readout of a SiPM-Based Pixelated Stilbene Array. Preprint (Version 1) Available at Research Square (2020)
231. X. Liang, J. Li, E. Antonecchia, Y. Ling, Z. Li, W. Xiao, Q. Chu, L. Wan, X. Hu, S. Han et al., NEMA-2008 and in-vivo animal and plant imaging performance of the large FOV preclinical digital PET/CT system discoverist 180. IEEE Trans. Radiat. Plasma Med Sci. **4**, 622–629 (2020)
232. J. Du, J.P. Schmall, K. Di, Y. Yang, P.A. Dokhale, K.S. Shah, S.R. Cherry, Performance comparison of different readouts for position-sensitive solid-state photomultiplier arrays. Biomed. Phys. Eng. Express **3**, 045019 (2017)
233. S. Sajedi, N. Zeraatkar, M. Taheri, S. Kaviani, H. Khanmohammadi, S. Sarkar, H. Sabet, M.R. Ay, Generic high resolution PET detector block using 12×12 SiPM array. Biomed. Phys. Eng. Express **4**, 035014 (2018)

Chapter 8
Dead Time Correction in Single-Photon Avalanche Diode Front Ends

8.1 Introduction

Dead time, i.e., the period during which an SPAD is not sensitive to incoming photons, arises from the inherent need to control the avalanche process [1, 2]. Specifically, when a carrier triggers an avalanche current, the current takes a short time to rise and it is then quenched, and typically, the device is further held in the quenched state for a while to mitigate afterpulsing. The SPAD is then reset. In each of these states, i.e., from the onset avalanche to a subsequent reset, the device is not able to register another photon. Therefore, the dead time is intricately related to the SPAD's operational characteristics.

Different methodologies for effectively controlling the avalanche process have been proposed. In one implementation called the time-gated mode [3], the SPAD is biased just below its breakdown level. The bias is transiently increased to a value above the breakdown voltage through the application of a calibrated voltage pulse. The device is held above breakdown for a short period of time corresponding to the width of the gating pulse, during which time an avalanche event may occur. The falling edge of the gate pulse lowers the voltage across the device and thus quenches any avalanche current. The time gate pulse width is kept short to limit the duration of the avalanche, usually to around a few nanoseconds. The resulting speed-up in the quench and reset operation improves the detection bandwidth and the active range of the device. The correct operation of the device, in the gated mode, requires precise timing of the gating pulse in order to match the photon arrival time. As a result, the utility of this approach is limited to applications wherein the arrival times are known.

In most applications, however, the arrival of the photon is not predetermined as there is no temporal information available regarding the photon source. Therefore, in this mode, which we term the free-running mode, the photon arrival cannot be predicted with any reasonable accuracy. Therefore, the onset of the avalanche needs to be dynamically sensed followed by immediate initiation of the quench and

M. Dandin et al., *Single-Photon Avalanche Diodes and Photon Counting Systems*,
https://doi.org/10.1007/978-3-031-64334-7_8

recovery operations. More importantly, in the free-running mode, a SPAD must be fast enough to temporally resolve individual photons absent in temporal information of the photon source. This means that ideally, from a system point of view, in a free-running SPAD device, the photon arrival rate must be linearly related to the output count rate.

In practice, the linearity between the output count rate and the photon arrival rate quickly breaks down as the photon rate of arrival increases: the SPAD's output signal saturates, and different photon arrival rates cannot be measured as the sensor's linearity is lost. This can be verified heuristically by measuring the time the sensor spends in its quench-and-reset cycle; the more time the detector spends in a state where it cannot detect additional incoming photons, the more dead time there is, and the more nonlinearity is observed as the photon arrival rate increases. Thus, in the free-running mode, high photon arrival rates can exacerbate the overall dead time, compromising linearity. Furthermore, even at low average photon arrival rates, because of shot noise, the instantaneous photon arrival rate can exceed the linear range of the detector, causing a nonlinear response.

Dead time in the free-running mode is further affected by current pulses that are temporally correlated with previous avalanche events (i.e., afterpulsing), by current pulses resulting from spurious emissions of electron-hole pairs due to thermal energy, and by current pulses originating from field-assisted generation from band-to-band tunneling.

In this chapter, we investigate the effect dead time in single-photon avalanche diode front ends from a system point of view. Specifically, we propose and validate a model for correcting measured count rates in SPAD front ends in the presence of dead time. The model is compared against the standard count rate correction model reported in the literature and validated experimentally using a pg-SPAD device.

We conclude by showing that the proposed count rate correction model is in better agreement with the true count rate of the SPAD, in contrast with the standard count rate correction model. We found that the proposed count rate correction model had a prediction error of about 2.1% relative to an 18.8% prediction error obtained from the standard count rate correction model typically used in the literature. Below, we present a theoretical framework describing the standard dead time correction model as well as the novel model. We then feature experimental results comparing count rate-corrected measurements using the two models.

8.2 Theoretical Framework

8.2.1 Standard Count Rate Correction Model

The arrival of photons on a detector's surface is a random process with an average rate of arrival λ, and it is a Poisson process. Consequently, the distribution of the photon inter-arrival times may be modeled as an exponential process. In non-paralyzable detection systems (i.e., in systems having a non-extending dead time), events arriving during the detection dead time (τ) are simply lost without altering the dead time τ. As such, the probability of detection may be expressed as the probability that an arbitrary inter-arrival time (X), for an incoming event rate (λ_{in}), is greater than the dead time τ of the detector. Given the exponential distribution of inter-arrival times, this probability is expressed as shown below in Equation 8.1:

$$P(X > \tau) = e^{-\lambda_{in} \cdot \tau} \tag{8.1}$$

Considering that the expected time between arrivals at the input of the detector is given by the inverse of the average arrival rate ($1/\lambda_{in}$), the expected inter-arrival time between detected events at the output of the detector may be characterized by an exponential distribution shifted by the dead time (τ). The expected time between measurements at the output is then

$$\frac{1}{\lambda_{out}} = \frac{1}{\lambda_{in}} + \tau \tag{8.2}$$

In other words, given a measured count rate (λ_{out}) and a detector dead time (τ), the true/actual count rate (λ_{in}) may be estimated by

$$\lambda_{in} \approx \frac{\lambda_{out}}{1 - (\lambda_{out}\, \tau)} \tag{8.3}$$

A corresponding expression may be equivalently expressed as

$$\lambda_{out} \approx \frac{\lambda_{in}}{1 + (\lambda_{in} \cdot \tau)} \tag{8.4}$$

Thus, we can calculate the probability of detection (P_m)as the ratio of output and input rates:

$$\frac{\lambda_{out}}{\lambda_{in}} = \frac{1}{1 + \lambda_{in} \cdot \tau} \tag{8.5}$$

Equation 8.4 is the standard model for estimating the output count rate (λ_{out}) as a function of the true count rate (λ_{in}) and the non-extending detector dead time (τ) [4]. This model has been extensively used in the SPAD literature to compensate for counts lost due to limited measurement bandwidth as well as for hold-off time compensation in photon counting applications.

Despite its popularity, the standard model does not provide a complete theoretical description of the underlying statistical processes giving rise to the measured count rate at the output of the detector. For instance, Equation 8.1 describes the probability that an arbitrary event inter-arrival time X is greater than the dead time interval τ. This implicitly assumes that an incoming photon arriving at the input of the detector at an arbitrary time t_2 is separated in time from the last detected event at time t_1 by one inter-arrival time window.

However, the probability of detecting the nth event/photon deviates from Equation 8.1 if the $(n-1)$th photon had arrived at the input of the detector during the dead time interval triggered by the detection of $(n-2)$th photon. In this case, the last detected event, corresponding to $(n-2)$th event at the output, would be separated from the nth photon at the input of the detector by two inter-arrival times. The operator X in Equation 8.1 now becomes 2(X) since the probability of detection for the nth photon now requires that the sum of two inter-arrival times (X) be greater than the dead time (τ).

Consequently, we must consider the probability of a sum of inter-arrival times to be greater than the detector's dead time. The sum of k exponentially distributed variables (x) having a common exponential distribution with a mean of $1/\lambda$ follows an Erlang distribution:

$$f(x) = \lambda \frac{(\lambda x)^{k-1}}{k-1} e^{-\lambda t} \tag{8.6}$$

Based on this distribution, the presence of dead time induces a shift in the probability distribution of events at the output of the detector. Such higher-order effects associated with varying statistics of dropped events within the dead time interval alter the expected time between measurements (Equation 8.2) and the probability of detection Equation 8.5 at the output of the detector. These second-order effects become critical as the expected number of event arrivals within the dead time $(\lambda_{in} \cdot \tau)$ increases. As such, the prediction of the standard model starts to degrade at higher input rates (λ_{in}) and/or for longer dead time intervals (τ).

8.2.2 *Proposed Count Rate Correction Model*

The proposed model is based on calculating the effective probability $P\,(n\text{th}\,)$ of detecting the nth photon by exploring all $(n-1)$ possibilities regarding the last detected event. Based on Equation 8.6, the probability that a time interval Y, made up of ***k*** inter-arrival times (i.e., $\mathrm{Y} = \mathrm{kX}$), is greater than the length of the dead time τ is described as

$$P(\mathrm{Y} > \tau) = \mathrm{e}^{-\lambda_{in}\cdot\tau} \cdot \sum_{0}^{k-1} \frac{(\lambda_{in} \cdot \tau)^{k}}{k!} \tag{8.7}$$

The summation index k ranges from 1 to $(n-1)$ and represents all the possible number of events that may occur within the dead time generated by the last detected event prior to the arrival of the nth photon. In other words it represents the possible number of inter-arrival periods separating the nth event at the input of the detector from the last detected event at the output of the detector. In other words, it characterizes the possible number of dropped events between the last detected event and the nth event currently registered by the detector.

The detection probability of the nth event can only be calculated with respect to the last detected event; this is inherent to the operation of any detector having a non-paralyzable, i.e., non-extending, dead time. However, as described above, the last detected event may correspond to any one of the $(n-1)$ previous events. This means that every dropped event improves the detection probability of the next subsequently arriving event, resulting in a widening of the probability distribution of event inter-arrival times at the output of the detector. Therefore, the detected events may be characterized with a higher probability of detection, depending of the mean arrival rate (λ_{in}) and the dead time (τ) as shown in Equation 8.7.

Here, for simplicity, we set the probability of detection of the first photon in a sequence of photons arriving on the detector to 1. The probability of detection for subsequent photons is calculated based on the probability of the last detected photon which could potentially be at any point behind the current photon. In case of a second photon arriving at the input, the last detected photon necessarily corresponds to the first photon. Therefore, the probability of detection for the second photon may be written as

$$P(2) = P(1) \cdot e^{-(\lambda_{\text{in}}\tau)} \tag{8.8}$$

For the third photon arriving at the input of the detector, the choice of the last detected photon may correspond to either the first or the second photon. The probability for the former case corresponds to the evaluation of Equation 8.7 at $k = 2$ multiplied by P(1). The probability expression for the latter case corresponds to evaluating Equation 8.7 at $k = 1$ multiplied by P(2). Therefore, the effective probability of detecting the third event/photon at the input of the detector P(n = 3) corresponds to the summation of the two probability expressions. Similarly, the last detected event relative to the fourth photon at the input may correspond to any one of the first, second, or third detected events at the output of the detector. Hence, calculating P(n = 4) involves the summation of all corresponding probabilities with regard to the last detected event. The effective probability of detecting the nth photon at the input of the detector may therefore be represented as shown in Equation 8.9, where x represents the various choices regarding the last detected event:

$$P\,(n\text{th}) = \sum_{x=1}^{n-1} P\,(n\text{th} \mid x) \times P(x) \tag{8.9}$$

Figure 8.1 shows the probability P (nth) n for different values of ($\lambda_{\text{in}} \cdot \tau$) and for different average photon arrival rates (λ_{in}) and a fixed value of dead time (τ) set to 1 μs. As observed, the effective probability of detection quickly settles to a constant value as n increases. Expressing Equation 8.9 as a function of the detector dead time (τ) and the average event arrival rate (λ_{in}) yields Equation 8.10:

$$P\,(nth) = \mathrm{e}^{-\lambda\tau} \times \sum_{k=1}^{n-1} \left(\frac{(\lambda_{\text{in}}\tau)^{k-1}}{(k-1)!} \times P(n-k) \times \prod_{x=n-1}^{n-k+1} 1 - P(x) \right) \tag{8.10}$$

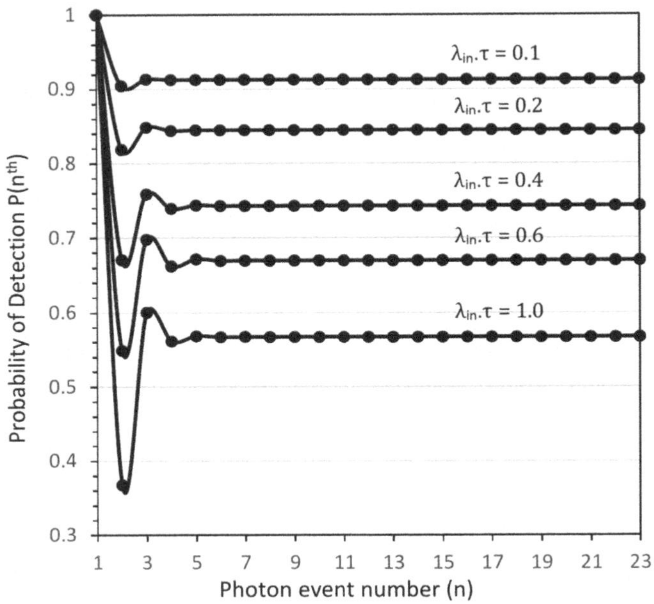

Fig. 8.1 P (nth) versus n for different $\lambda_{in} \cdot \tau$ values corresponding to a τ of 1μ s and λ_{in} values ranging from 100 KHz to 1 MHz

8.3 Experimental Methods

8.3.1 Perimeter-Gated Single-Photon Avalanche Diode and Equivalent Circuit Model

Our dead time evaluation test bed included a perimeter-gated single-photon avalanche diode (pg-SPAD) monolithically integrated with a mixed-signal front-end readout and processing circuit. As explained throughout this book, the pg-SPAD is an avalanche diode that exhibits a variable dark count rate, depending on the voltage on its perimeter gate [5–8]. The pg-SPAD allowed known average dark count rates to be obtained simply by setting the perimeter gate voltage to a value corresponding a desired average dark count rate. The front-end circuit was designed to introduce a known delay in the readout path; this provided the ability to introduce known dead times.

As such, being able to set the dark count rate and the delay in the readout path, the pg-SPAD was found to be an ideal platform for evaluating the novel count rate correction model. Specifically, our experiments consisted of first recording the true dark count rate, i.e., the measured dark count rate with no delay and for a specific gate voltage. We then compared the measured true dark count rate with the corrected dark count rates obtained from both the standard model and the proposed model, in both cases with known delays in the readout path. One limitation of our approach, however, is that we are using the device in the dark, simply using dark counts to model the arrival process. Despite this approach, the emission of these carriers still follows a Poisson process and thus an equivalent substitute to a well-controlled photon source.

The pg-SPAD and the front-end circuit were fabricated in a $0.5 - \mu m$, 3-metal, 2-poly, single-well CMOS process. The front-end circuit not only provided an interface to the pg-SPAD, but it also served to quench and reset the diode upon avalanche; furthermore, the front-end circuit was also used to generate a digital output pulse whose leading edge corresponded in time with the onset of avalanche. In order to estimate a count rate, a train of output pulses was counted in software during a specified monitoring period.

The pg-SPAD's top and cross-sectional views are shown in Fig. 8.2a; the device is a p-n junction equipped with a polysilicon gate disposed at the periphery of the junction's active region. As previously reported, perimeter gating is an effective means of suppressing premature edge breakdown, which allows high signal-to-noise ratio (SNR) Geiger-mode operation, while also providing the ability to actively tune the diode's breakdown voltage. During operation, when a large-magnitude negative voltage is applied to the polysilicon gate, the region of maximum charge generation shifts from the edge and surface of the diode to the metallurgical junction, which ensures Geiger-mode operation with large SNR.

Figure 8.2b shows the pg-SPAD's operation in a passive quenching and reset configuration (inset). The device is reversed-biased between a high-voltage V_{DD}(e.g. + 5 V) and a low voltage V_{Low} or V_A (e.g., −13 V). The circuit is quenched using a large quenching resistor $R_C(\sim 100\,k\Omega)$; a small termination resistor $R_A(\sim 50\Omega)$ may also be used at the anode. The cathode voltage (V_{CATH}) is read out as the output voltage pulse V_{out}. The pulse includes two periods at which the device is at quiescence, denoted I and II; between these two periods, the device avalanches,

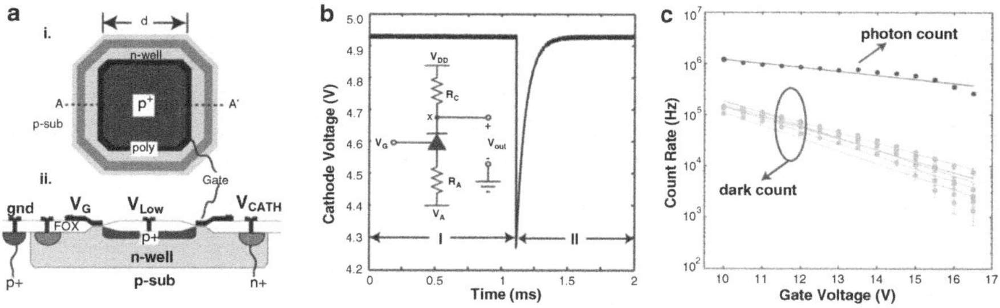

Fig. 8.2 (**a**) Top view (i) and cross-sectional view (ii) of a perimeter-gated single-photon avalanche diode (pg-SPAD). The device consists of a p+/n-well junction fabricated in a p-substrate. A polysilicon gate is placed at the device's periphery. The p-substrate is held at ground to reverse bias the parasitic p+/n-well junction. (**c**) Performance of the pg-SPAD as a function of gate voltage showing the count rate measured for different gate voltages under illumination and in the dark. The dark count rate data are shown for multiple pg-SPADs located on the same chip

marked by the sharp drop in voltage, and the device is reset as indicated by the relatively long exponential transient towards the quiescent level of period II.

As shown in Fig. 8.2b, the temporal dynamics associated with the passive quenching mode can be in the sub-millisecond regime. For faster dynamics, active quenching may be used; specifically, a transistor can be used to reset the cathode node towards V_{DD} when an avalanche occurs, thereby accelerating the transition towards the quiescent level of period II. A transistor can also be used to further accelerate the quenching phase at the onset of avalanche.

Figure 8.2c shows the pg-SPAD's performance when an active quenching and reset readout scheme is used. The data also show the effect of the perimeter gate voltage on the dark count rate and on the photon count rate (i.e., under illumination). For increasing perimeter gate voltage magnitudes, the dark count decreases; a relatively smaller decrease is also observed in the photon count rate; however, a high SNR can be achieved, which maximizes sensitivity to weak photon fluxes [7].

The pg-SPAD's electrical operation can be best understood by analyzing the first-order SPAD equivalent circuit model shown in Fig. 8.3. The model consists of an R-C circuit that simulates the avalanche dynamics of the pg-SPAD for a fixed gate voltage. The model can be used in either passive, active, or mixed (active/passive) modes for quenching and reset.

It utilizes a set of voltage-controlled switches (S1–S4) to simulate an avalanche in the pg-SPAD. A stimulus is provided at the input terminal of the switch S3. The stimulus is a pulse train whose rising edges are used to simulate photon arrival times; thus, pulse widths and amplitudes are not relevant as only the rising edges trigger the circuit. For example, when the voltage level at the cathode terminal is

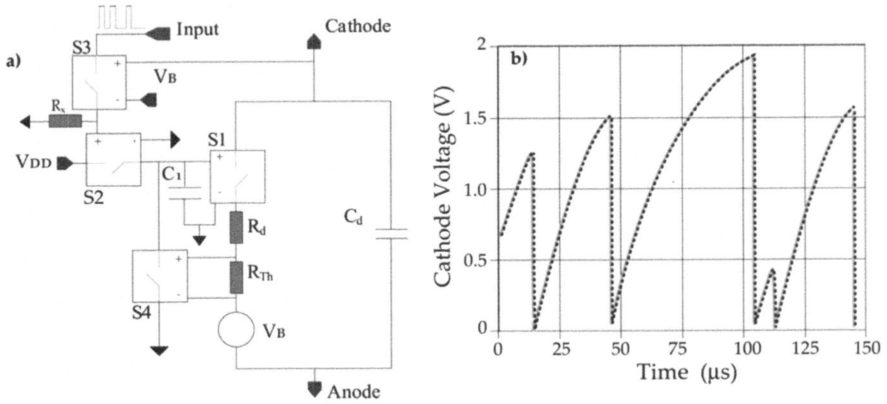

Fig. 8.3 (**a**) Equivalent circuit model of a pg-SPAD for a fixed gate voltage. The circuit was implemented in Cadence using a set of voltage-controlled switches (S1, S2, S3, and S4), capacitors C1 and Cd, and resistors R_{Th}, R_d, and R_x. The input is provided as a random pulse train simulating photon arrival. The breakdown voltage is modeled as a DC voltage source set at V_B. (**b**) Simulated Geiger-mode events (dashed line) utilizing the equivalent circuit model compared with measured data from a fabricated pg-SPAD (solid line). The pg-SPAD was operated in passive quenching mode both in simulation and during actual operation

at or above the device's breakdown voltage and a rising edge in the stimulus signal occurs, switches S3 and S2 are closed, which also triggers S1 to close.

These actions represent an avalanche event because when these switches are closed, a path to ground is provided, and the diode's reverse bias capacitance C_d discharges. Thus, the circuit simulates the device going from a quiescent state to an avalanche state, as shown in period I above. We note that C_d can be assigned the value of the measured reverse-biased junction capacitance of the diode at a fixed reverse-bias voltage. However, better agreement between the model and measured data is observed when C_d is modeled behaviorally as a voltage-dependent capacitance, as is the case in the actual device. (See Refs. [9, 10], for example, behavioral models of a SPAD's reverse-biased junction capacitance.)

The capacitor C1 is used to ensure that the avalanche operation does not end in response to the stimulus' falling edge, which causes the switch S2 to open. Quenching does not occur until the discharge current falls below a self-sustaining current threshold. This self-sustaining current level is known as the latching current [11, 12], and an avalanche is improbable when the diode current is below this level. In our model, the latching current level is detected across R_{th}, and when it is reached, S4 is closed and C_1 discharges; this causes S1 to open, and the avalanche current is turned off, which simulates quenching.

At that point, the capacitor C_d recharges exponentially towards V_{DD}, which marks the transition from the quenched state to the quiescence level of period II. The equivalent circuit model's validity was verified by comparing the measured waveform of a test device biased at 2 V above its breakdown and passively quenched with a 100-kΩ ballast resistor. As shown in Fig. 8.3b, the equivalent circuit model's output (dashed line) shows good agreement with measured results (solid line).

8.3.2 Front-End Interface and Readout Circuit for Studying Dead Time

The SPAD device model described above was used to simulate a front-end processing and readout as shown in Fig. 8.4a.

The circuit was designed to provide rapid quenching and rapid reset in response to an avalanche in the pg-SPAD. Furthermore, the circuit was also designed to provide an adjustable hold-off time. The hold-off time generation was achieved using a hold-off time generation circuit and an adjustable output pulse width circuit, both of which are described in the following subsections and in Fig. 8.4a and b.

8.3.3 Active Quenching and Active Reset Functionalities

The front-end circuit includes a transistor M5 that pulls up the cathode of the pg-SPAD when a control signal is at logic 0; this control signal is thus the reset signal.

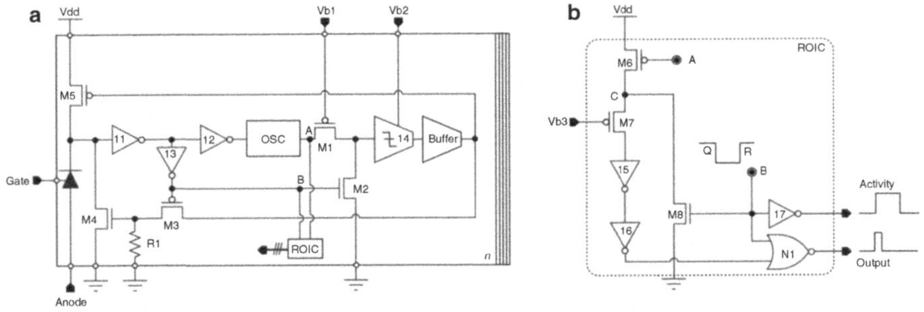

Fig. 8.4 (**a**) Circuit architecture of a pixel array where each pixel is equipped with a pg-SPAD and a front-end processing and readout circuit. The array is an **n** × **n** array where each pixel shares power and ground, analog bias voltages, and gate and anode bias voltages. (**b**) Circuit architecture of the readout and interface circuit (ROIC) included in each pixel

Conversely, the transistor M4 pulls the cathode towards ground, when the voltage on its gate is a logic 1, i.e., when the quenching signal is asserted. The period between quenching and reset can be programmed by setting the bias voltage Vb1, which controls the channel current of M1. Specifically, the oscillator (OSC) in Fig. 8.4a generates a periodic pulse train used to charge a capacitive node (integration node C) following the detection of an avalanche event, i.e., when the cathode voltage drops below the switching threshold of inverter I1. As such, the voltage Vb1 controls the channel current of M1, which in turn controls the rate at which the integration node C charges.

For example, when the integration node C is charged at full capacity, the end of the hold-off time is reached, and the output voltage of the comparator I4 switches and is then buffered to generate the control signal that turns on M5 and pulls the cathode towards V_{DD}. At that point, the pg-SPAD is reset. As such, the integration node, the comparator I4, and the transistor M2 act together as an analog counter that tallies the cycles of the pulse train generated by the oscillator. Additional control of the hold-off time may be achieved using Vb2, which can be used to alter the switching the threshold of the comparator I4. The rising cathode voltage causes the output of I1 to drop, deactivating the oscillator signal and resetting the integration node to zero. The control signal then switched to logic 1 and M5 is turned off. The SPAD device is now armed for the next detection event.

A wide range of variable hold-off times, hence dead times, can be generated by coarse and fine adjustments through V_{bias1} and V_{bias2}, respectively. V_{bias1} controls the charging duty cycle through transistor M1, which determines the slew rate at the integration node. V_{bias2} determines the signal amplitude that is required to trigger the termination of quench and the initiation of reset operation.

Figure 8.5 illustrates experimental measurements of the hold-off time (SPAD activity pulse width) for the SPAD test unit described above. The hold-off time is measured for a set of bias values with V_{bias1} ranging from 2.9 to 3.5 V and V_{bias2} ranging from 3.0 to 5.0 V.

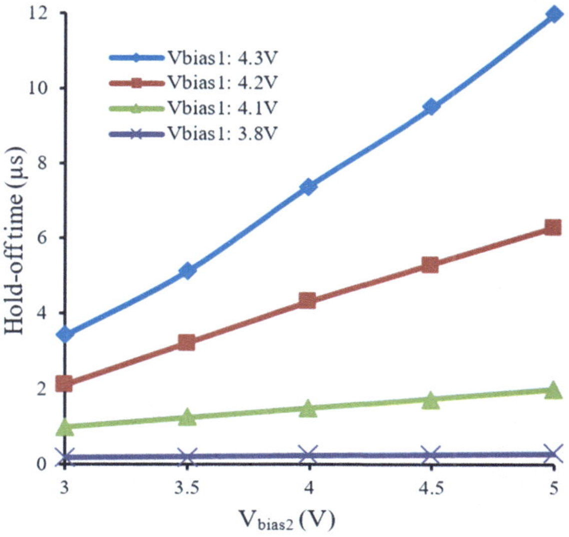

Fig. 8.5 Measured hold-off times, i.e., dead times, for a set of Vbias1 and Vbias2 values

8.3.4 Adjustable Output Pulse Width Generation Circuit

In typical SPAD pixels the width of the output pulse generated by the pixel front-end electronics is defined by the operational duty cycle of the SPAD device. Consequently, in SPAD array architectures involving a common readout channel such as in SiPMs or in SPAD imagers, the performance of the detection system may become constrained by the activity of the individual pixels.

For example, the duration of the coincidence window in SiPM detectors is statically defined by the duration of individual pixel pulse responses. Similarly, in SPAD-based imagers and photon counters, the effective throughput of the detector becomes a function of device parameters such as the operational duty cycle of an individual SPAD device. As such, there is a need to decouple device and system-level parameters. So, we propose a SPAD device readout interface capable of controlling the SPAD pixel output pulse width independently of the device operational duty cycle. Thus, our proposed front-end system (Fig. 8.4) optimizes the detection performance of the device by controlling the timing associated with the SPAD avalanche-quench-reset cycle while independently allowing for adjusting of the SPAD pixel output pulse width.

The generation of the output pulse is initiated by the onset of the quenching action at the cathode terminal of the SPAD device and terminated at the end of a hold-off delay period set by V_{bias3} parameter. The variable output pulse width is implemented by the same mechanism as used to vary the hold-off time. At quiescent, the high voltage at the SPAD cathode terminal causes the output of the NOR gate in Fig. 8.4b to be low. Following detection of an event, the quenching action lowers the voltage on node B, causing the output of the NOR gate to go high, hence triggering the rising edge of the output pulse.

The concurrent deactivation of M8 and activation of the oscillatory signal at node A in parallel with the initiation of the output pulse subsequently leads to the activation of node C which turns off the output of the NOR gate and terminates the output pulse. The resetting of the NOR gate output occurs after a delay interval set by the value of V_{bias3}. If the value of V_{bias3} is set too high, the trailing edge of the output pulse will coincide with the reset operation of the SPAD device in which case the width of the SPAD pixel output pulse will correspond to the hold-off time of the SPAD device (width of the SPAD activity pulse).

However, the value of V_{bias3} can be set such that the input node C of the NOR gate is activated prior to the initiation of the SPAD device reset operation. This will result in an output pulse that is an adjustable fraction of the period associated with the SPAD activity pulse. The proposed readout interface enables the detector to operate at a reduced duty cycle relative to the operational duty cycle of the SPAD detector, even at high activity rates. This is in contrast with previous approaches in which the SPAD activity cycle directly controls the duty cycle of the detector output signal.

8.4 Measured Results

As a test of its predictive accuracy, the proposed model was used to reconstruct the mean DCR of the pg-SPAD test device at various DCR levels. The triggering rate of the SPAD device, corresponding to the magnitude of the DCR in the test device, was altered by varying the bias level applied to a perimeter field gate [5, 6]. The dead-time free DCR, i.e., with no hold-off generated, was measured using a passive quenching set up at three different gate bias levels, using two different measurement schemes. In the first scheme, the signal at the cathode terminal of the test device was directly recorded within multiple data frames using a high-speed digitizer. A Matlab program was then used to count the signal transitions representing dark events. Since no hold-off was associated with this measurement process, the measured count rates were used as control data.

The process was repeated using a fixed dead time of 10.58 μs into the measurement process. The second set of measured rates represents the dead time-corrupted data. We then attempted to reconstruct the true count rate by applying the standard and proposed models separately to the corrupted data set.

The prediction error is significantly lower for the proposed model than for the standard correction model. Our comparison results are presented in Fig. 8.6. At an average count rate of 12k, both models are relatively accurate, but as the count rate is increased, the discrepancies grow. At a count rate 48 kHz, the standard model exhibits a prediction error of 18.8% compared to 2.1% for the proposed model. At a count rate of 117 KHz, the prediction error of the standard model and the proposed model increases to 40% and 2.6%, respectively.

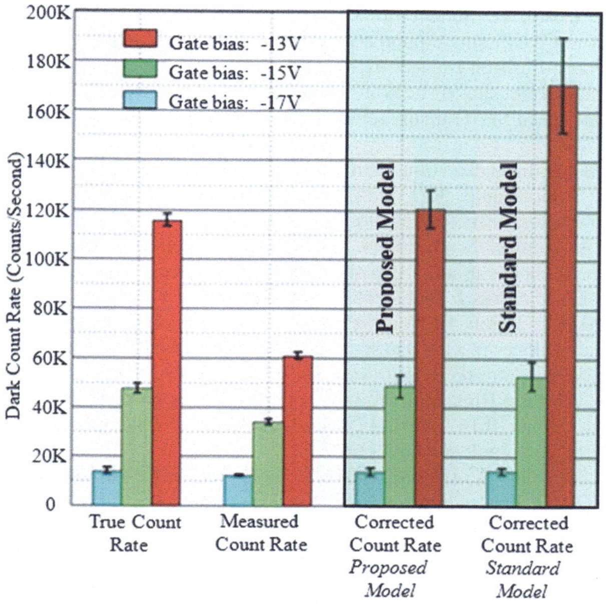

Fig. 8.6 Performance comparison between the proposed and the standard correction model

8.5 Conclusions

In this chapter, we reported on an analytical model for count rate correction in SPAD front ends due to dead time-related losses. We proposed a new correction model that can be used in the back end to complement the operation of high-speed photon counting hardware. The new model enhances the dynamic range of the device by compensating for the adverse effect of dead time at high count rates.

Furthermore, the chapter also featured the design of a fast and compact front-end system for high-bandwidth SPAD operation in a standard CMOS process. We also presented a novel method for linear hold-off time generation. Additionally, we featured a novel structure for event pulse generation as part of the front-end electronics interfaced with the pg-SPAD.

References

1. L. Neri, S. Tudisco, F. Musumeci, A. Scordino, G. Fallica, M. Mazzillo, M. Zimbone, Note: dead time causes and correction method for single photon avalanche diode devices. Rev. Sci. Instrum. **81**(8), 086102 (2010) [Online]. Available: http://aip.scitation.org/doi/10.1063/1.3476317

2. S. Chick, R. Coath, R. Sellahewa, R. Turchetta, T. Leitner, A. Fenigstein, Dead time compensation in CMOS single photon avalanche diodes with active quenching and external reset. IEEE Trans. Electron Devices **61**(8), 2725–2731 (2014)
3. K. Morimoto, A. Ardelean, M.-L. Wu, A.C. Ulku, I.M. Antolovic, C. Bruschini, E. Charbon, Megapixel time-gated SPAD image sensor for 2D and 3D imaging applications. Optica **7**(4), 346 (2020)
4. J.W. Müller, Generalized dead times. Nucl. Instrum. Methods Phys. Res. Sect. A: Accel. Spectrom. Detect. Assoc. Equip. **301**(3), 543–551 (1991) [Online]. Available: https://linkinghub.elsevier.com/retrieve/pii/016890029190021H
5. M. Dandin, A. Akturk, B. Nouri, N. Goldsman, P. Abshire, Characterization of single-photon avalanche diodes in a 0.5 μm standard CMOS process–Part 1: Perimeter breakdown suppression. IEEE Sensors J. **10**(11), 1682–1690 (2010)
6. M. Dandin, M.H.U. Habib, B. Nouri, P. Abshire, N. McFarlane, Characterization of single-photon avalanche diodes in a 0.5 μm standard CMOS process–Part 2: Equivalent circuit model and Geiger mode readout. IEEE Sensors J. **16**(9), 3075–3083 (2016)
7. M. Dandin, P. Abshire, High signal-to-noise ratio avalanche photodiodes with perimeter field gate and active readout. IEEE Electron Device Lett. **33**(4), 570–572 (2012) [Online]. Available: http://ieeexplore.ieee.org/document/6165634/
8. B. Nouri, M. Dandin, P. Abshire, Characterization of single-photon avalanche diodes in standard CMOS, in *Proceedings of IEEE Sensors*, (2009). https://doi.org/10.1109/ICSENS.2009.5398384. ISBN 9781424445486
9. A. Dalla Mora, A. Tosi, S. Tisa, F. Zappa, Single-photon avalanche diode model for circuit simulations. IEEE Photon. Technol. Lett. **19**(23), 1922–1924 (2007)
10. F. Zappa, A. Tosi, A.D. Mora, S. Tisa, SPICE modeling of single photon avalanche diodes. Sensors Actuators A: Phys. **153**(2), 197–204 (2009)
11. S. Cova, M. Ghioni, A. Lacaita, C. Samori, F. Zappa, Avalanche photodiodes and quenching circuits for single-photon detection. Appl. Opt. **35**(12), 1956 (1996) [Online]. Available: https://www.osapublishing.org/abstract.cfm?URI=ao-35-12-1956
12. R.H. Haitz, Model for the electrical behavior of a microplasma. J. Appl. Phys. **35**(5), 1370–1376 (1964)

Chapter 9
Conclusions, Contributions, and Future Work

This book has focused on various aspects of single-photon avalanche (SPAD) technology, particularly centering on perimeter gating and its performance in standard CMOS processes. Although the state of the art is dominated by traditional CMOS SPAD architectures, the experimental results shown herein set perimeter-gated SPADs as viable substitutes for these architectures and even as an alternative that confers additional functionality. For example, we have shown that perimeter readily confers the ability to alter breakdown voltages, allowing one to tighten breakdown voltage distribution in a large array by equalizing breakdown voltages across different SPAD pixels. Furthermore, we have shown that perimeter gating may also be used in a hardware security framework to leverage the underlying physically unclonable function of a pg-SPAD imager to create chip-specific signatures.

These are but a few advantages conferred by perimeter gating. We posit that techniques that actively modulate the avalanche region spatially, like perimeter gating, can confer significant advantages heretofore unforeseen in the SPAD literature. For example, with perimeter-gated microcells in an SiPM, it is possible to locally impede the probability of avalanche in order to quench overly active microcells. This is a well-sought-after feature in SiPMs, especially for high-energy physics applications.

Below, we summarize some of the most salient findings featured in the book, from Chaps. 2 to 8. We do so in order to delineate the metes and bounds of perimeter-gated SPADs relative to the state of the art. In doing so, we specifically point out our key contributions to detector design, readout circuits, and novel applications that utilize SPADs.

M. Dandin et al., *Single-Photon Avalanche Diodes and Photon Counting Systems*,
https://doi.org/10.1007/978-3-031-64334-7_9

9.1 Conclusions and Contributions

Chapter 2 focused on the modeling and empirical characterization of pg-SPADs in standard CMOS processes. We showed that perimeter gating caused a spatial shift in the multiplication region; with perimeter gating, the multiplication region moved away from the diode's surface and edges towards the metallurgical volumetric junction. This was confirmed using a calibrated and self-consistently solved device-physics model of the pg-SPADs. Empirically, the effects are perimeter gating can be seen as a shift in the breakdown voltage in the diode's reverse I-V characteristics.

Chapter 2 also discussed pg-SPAD modeling. We showed an equivalent circuit model that had an intrinsic mechanism that accounts for perimeter gating. Specifically, the breakdown voltage of the diode was modeled using a voltage-dependent voltage source, and the control parameter of that source was established from measured experimental data. We further showed an active quenching and an active reset circuit interfaced with a monolithically integrated pg-SPAD.

A key result of Chap. 2, beyond showing that perimeter gating is an effective means of curtailing PEB, is the reduction in count rate variance beyond the point of perimeter breakdown mitigation. From this result, we posited that perimeter gating can be used beyond the point of PEB mitigation to provide additional benefits. This hypothesis was supported by subsequently measured data, which suggest that further gains in DCR reduction are possible beyond the point of PEB suppression.

In Chap. 3, we explored the optoelectronic response of pg-SPAD. At first, we characterized a pg-SPAD's SNR as a function of monochromatic light intensity. We showed that SNR could be improved with perimeter gating, as the device's NEP decreased with increased perimeter gate voltage. This result confirmed that perimeter gating could be effected without compromising the device's ability to detect photons.

Having confirmed that, we studied the spectral responsivity of the diode as a function of perimeter gating. The results revealed a strong dependence of the spectral profile on perimeter gate voltage, but this was only in a relatively low perimeter gate voltage regime. Beyond PEB suppression, increasing the perimeter gate voltage did not result in significant changes in near-breakdown spectral responsivity. This further confirmed perimeter gating as a viable means for making high-performance SPADs in standard CMOS processes.

Next, in Chap. 3, we demonstrated a pg-SPAD imager, and we studied its pixel spectral response as well as its photo-response nonuniformity (PRNU). We found, as before, that perimeter gating helps regularize the device's spectral response and that the response was largely insensitive to perimeter gate voltage in the high perimeter gate voltage regime. Further, we found that the uncertainty in the PRNU measured was significantly decreased in the high perimeter gate voltage region. This latter result implies that excess noise in SPAD imagers may make the estimation of PRNU difficult. Perimeter gating thus provides a way to mitigate noise processes and facilitate PRNU measurements.

Chapter 4 focused in greater detail on the imager introduced in Chap. 3, presenting results on the effects of perimeter gating on dark signal nonuniformity (DSNU) and on PRNU. Furthermore, Chap. 4 characterized the imager performance in image acquisition tasks as a function of perimeter gating. Chapter 4 results show that both the DSNU and the PRNU can be modulated using perimeter gating and that any remaining fixed pattern noise (FPN) can be removed using image processing algorithms on the back end.

Chapter 5 establishes the use of perimeter gating as a means to create a hardware security layer inherently accessible by an image sensor's hardware. In the chapter, we show that the ability to modulate the DSNU using perimeter gating can be leveraged as a means to access several physically unclonable functions (PUFs) intrinsic to the imager. These PUFs can then be used to generate signatures that are camera-specific, thus providing trusted provenance information that may be embedded or included as metadata in an image taken by the imager. This work shows that pg-SPADs can be used as hardware security primitives.

In Chap. 6, we describe the silicon photomultipliers (SiPM), and we show that pg-SPADs can also be used to make these detectors. We show that digital or analog SiPMs can easily be implemented with pg-SPADs and further that reduced deadtime pixels may be implemented. Further, we show that perimeter gating may be used as a means to equalize dark count rates across an SiPM pg-SPAD array.

Chapter 7 surveys various readout strategies for SiPMs. We specifically focus on amplifier front ends that are typically encountered in nuclear imaging applications where an imager is made from multiple SiPMs, each SiPM consisting of a pixel of the imager. The design tradeoffs of these front ends were discussed, and several implementations were featured.

Chapter 8 proposed a novel count rate correction model for use in SPAD front ends. The model builds on the canonical approach to correct rates affected by dead time to provide a new way to estimating a wider range of possible events dropped within the dead time. We showed that pg-SPADs equipped with a hold-off delay generator could be used to evaluate the new count rate correction model, as perimeter gating and the delay generator could be used cooperatively to introduce desired count rates and delay times.

9.2 Future Work

This book discussed a particular implementation of SPADs and SPAD-based devices. Specifically, we showed perimeter gating and pg-SPADs as a platform technology that may be incorporated in SiPMs, imagers, and even as the elements of a hardware security layer inherent to an image sensor. The foreseeable future is an exciting one for pg-SPAD technology!

For example, developing more compact devices, focusing on pixel fill factor increases, and improving frame rates, all the while providing local perimeter gate control, are all areas of active research within our respective research groups. Furthermore, developing and demonstrating new use cases that leverage the benefits of perimeter gating are also on our short-term horizon. For instance, applications in visible light communication, hardware security, and applications in quantum computing are all being contemplated in our labs. In closing, it is our hope that this monograph will inspire additional developments in the field.

Index

M. Dandin et al., *Single-Photon Avalanche Diodes and Photon Counting Systems*,
https://doi.org/10.1007/978-3-031-64334-7

The manufacturer's authorised representative in the EU is Springer Nature Customer Service Centre GmbH, Europaplatz 3, 69115 Heidelberg, Germany. If you have any concerns regarding our products, please contact ProductSafety@springernature.com

Printed and bound by CPI Group (UK) Ltd, Croydon, CR0 4YY
01/10/2026
02225961-0003